Matrix Calculus

and

Kronecker Product

with

Applications

and

C++ Programs

Matrix Calculus
and
Kronecker Product
with
Applications
and
C++ Programs

Willi-Hans Steeb

International School of Scientific Computing,
Rand Afrikaans University

in collaboration with

Tan Kiat Shi

National University of Singapore

Published by

World Scientific Publishing Co. Pte. Ltd.

P O Box 128, Farrer Road, Singapore 912805

USA office: Suite 1B, 1060 Main Street, River Edge, NJ 07661

UK office: 57 Shelton Street, Covent Garden, London WC2H 9HE

Library of Congress Cataloging-in-Publication Data

Steeb, W.-H.

Matrix calculus and Kronecker product with applications and C++ programs / Willi-Hans Steeb, in collaboration with Tan Kiat Shi.

p. cm.

Includes bibliographical references and indexes.

ISBN 9810232411

1. Matrices. 2. Kronecker products. 3. Matrices -- Data processing. 4. Kronecker products -- Data processing. 5. C++ (Computer program language) I. Shi, Tan Kiat. II. Title.

QA188.S663 1997

530.15'29434--dc21 97-26420 CIP

British Library Cataloguing-in-Publication Data

A catalogue record for this book is available from the British Library.

This book is printed on acid-free paper.

Printed in Singapore by Uto-Print

Preface

The Kronecker product of matrices plays an important role in mathematics and in applications found in theoretical physics. Such applications are signal processing where the Fourier and Hadamard matrices play the central role. In group theory and matrix representation theory the Kronecker product also comes into play. In statistical mechanics we apply the Kronecker product in the calculation of the partition function and free energy of spin and Fermi systems. Furthermore the spectral theorem for finite dimensional hermitian matrices can be formulated using the Kronecker product. The so-called quantum groups rely heavily on the Kronecker product. Most books on linear algebra and matrix theory investigate the Kronecker product only superficially. This book gives a comprehensive introduction to the Kronecker product of matrices together with its software implementation in C++ using an object-oriented design.

In chapter 1 we give a comprehensive introduction into matrix algebra. The basic definitions and notations are given in section 1.1. The trace and determinant of square matrices are introduced and their properties are discussed in section 1.2. The eigenvalue problem plays a central role in physics. Section 1.3 is devoted to this problem. Projection matrices and projection operators are important in Hilbert space theory and quantum mechanics. They are also used in group theoretical reduction in finite group theory. Section 1.4 discusses these matrices. In signal processing Fourier and Hadamard matrices play a central role for the fast Fourier transform and fast Hadamard transform, respectively. Section 1.5 is devoted to these matrices. Transformations of matrices are described in section 1.6. The invariance of the trance and determinant are also discussed. Finite groups can be represented as permutation matrices. These matrices are investigated in section 1.7. The vec operator describes an important connection between matrices and vectors. This operator is also important in connection with the Kronecker product. Section 1.8 introduces this operator. The different vector and matrix norms are defined in section 1.9. The relationships between the different norms are explained. We also describe the connection with the eigenvalues of the matrices. The exponential function of a square matrix is useful in many applications, for example Lie groups, Lie transformation groups and for the solution of systems of ordinary differential equations. Sequences of vectors and matrices are introduced in section 1.10 and in particular the exponential function is discussed. Groups are studied in section 1.11. A number of their properties are given. Section 1.12 introduces Lie algebras. Applications in quantum theory are given in section 1.13. We assume that the Hamilton operator is given by a Hermitian matrix. We investigate the time evolution of the wave function (Schrödinger equation) and the time evolution of a matrix (Heisenberg equation of motion).

Sections 2.1 to 2.3 in chapter 2 give an introduction to the Kronecker product. In particular, the connection with matrix multiplication is discussed. In Section 2.4 permutation matrices are discussed. Section 2.5 is devoted to the trace and determinant of a matrix and their relation to the Kronecker product. The eigenvalue problem is studied in section 2.6. We calculate the eigenvalues and eigenvectors of Kronecker products of matrices. In order to investigate the spectral representation of Hermitian matrices we introduce projection matrices in section 2.7. Section 2.8 describes the spectral representation of Hermitian matrices using the Kronecker product. Fourier and Hadamard matrices are important in spectral analysis, such as fast Fourier transforms. These matrices are introduced in section 2.9 and their connection with the Kronecker product is described.The direct sum and the Kronecker product are studied in section 2.10. Section 2.11 is devoted to the vec-operator and its connection with the Kronecker product. Groups and the Kronecker product are investigated in Sections 2.12 and 2.13. In particular the matrix representation of groups is described. The inversion of partitioned matrices is discussed in section 2.14.

In chapter 3 we study applications in statistical mechanics, quantum mechanics, Lax representation and signal processing for the Kronecker product. First we introduce Pauli spin matrices and give some applications in section 3.1. The eigenvalue problem of the two point Heisenberg model is solved in detail. The one-dimensional Ising model is solved in section 3.2. Fermi systems are studied in section 3.3. We then study the dimer problem in section 3.4. The two-dimensional Ising model is solved in section 3.5. In section 3.6 the one-dimensional Heisenberg model is discussed applying the famous Yang-Baxter relation. Quantum groups are discussed in section 3.7. Section 3.8 describes the connection of the Kronecker product with the Lax representation for ordinary differential equations. Signal processing and the Kronecker product is discussed in section 3.9.

The tensor product can be considered as an extension of the Kronecker product to infinite dimensions. Chapter 4 gives an introduction into the tensor product and some applications. The Hilbert space is introduced in section 4.1 and the tensor product in section 4.2. Sections 4.3 and 4.4 give two applications. In the fist one we consider a spin-orbit system and the second one a Bose-spin system. For the interpretation of quantum mechanics (system and the measuring apparatus) the tensor product and Kronecker product is of central importance. We describe this connection in section 4.5.

In chapter 5 the software implementations are given. An object-oriented design is used. A vector and matrix class (abstract data types) are introduced together with all the operations, i.e. trace, determinat, matrix multiplication, Kronecker product etc.. The classes are written on a template basis so that every appropiate data type can be used with these classes, for example, integer, double, verylong, rational, symbolic. A number of applications are given.

In most sections a large number of examples and problems serve to illustrate the mathematical tools. The end of a proof is indicated by ♠. The end of an example is indicated by ♣.

If you have any comments about the book please send them to us at:

email: WHS@RAU3.RAU.AC.ZA

Web page: http://zeus.rau.ac.za/steeb/steeb.html

Acknowledgements

I am grateful to John and Catharine Thompson for careful proof-reading of the manuscript and checking a number of results. I also thank Dr. Fritz Solms for discussions on object-oriented programming and C++. Finally, I appreciate the help of the Lady from Shanghai Zhao Gui Zhu.

Contents

Symbol Index

$\mathbf{Z}$	the set of integers
$\mathbf{N}$	the set of positive integers: natural numbers
$\mathbf{Q}$	the set of rational numbers
$\mathbf{R}$	the set of real numbers
$\mathbf{R}^+$	non negative real numbers
$\mathbf{C}$	the set of complex numbers
$\mathbf{R}^n$	the n dimensional real linear space
$\mathbf{C}^n$	the n dimensional complex linear space
i	$:= \sqrt{-1}$
$\mathbf{u}, \mathbf{v}$	column vectors in $\mathbf{C}^n$
$\mathbf{0}$	zero vector in $\mathbf{C}^n$
$\otimes$	Kronecker product (direct product, tensor product)
$\oplus$	direct sum
$\|\cdot\|$	norm
$(\, , \,)$	scalar product (inner product) in $\mathbf{C}^n$
tr	trace
det	determinant
I	unit matrix (identity matrix)
A^T	transpose of the matrix A
$\boldsymbol{\sigma}$	permutation
P	permutation matrix
U	unitary matrix
Π	projection matrix
$[\, , \,]$	commutator
$[\, , \,]_+$	anti-commutator
δ_{jk}	Kronecker delta 1 if $j = k$ and 0 if $j \neq k$
λ	eigenvalue
G	group
σ	spin
$\sigma_x, \sigma_y, \sigma_z$	Pauli spin matrices
$c^\dagger, c$	Fermi creation and annihilation operators
$b^\dagger, b$	Bose creation and annihilation operators
T	transfer matrix
$\hat{H}$	Hamilton operator
N	Number of lattice sites
J	exchange constant
$Z(\beta)$	partition function

Symbol Index

Chapter 1

Matrix Calculus

1.1 Definitions and Notation

We assume that the reader is familiar with some basic terms in linear algebra such as vector spaces, linearly dependent vectors, matrix addition and matrix multiplication.

Throughout we consider matrices over the field of complex numbers $\mathbf{C}$ or real number $\mathbf{R}$. Let $z \in \mathbf{C}$ with $z = x + iy$ and $x, y \in \mathbf{R}$. Then $\bar{z} = x - iy$. In some cases we restrict the underlying field to the real numbers $\mathbf{R}$. The matrices are denoted by A, B, C, D, X, Y. The matrix elements (entries) of the matrix A are denoted by a_{jk}. For the column vectors we write $\mathbf{u}$, $\mathbf{v}$, $\mathbf{w}$. The zero column vector is denoted by $\mathbf{0}$. Let A be a matrix. Then A^T denotes the transpose and $\bar{A}$ is the complex conjugate matrix. We call A^* the adjoint matrix, where $A^* := \bar{A}^T$. A special role is played by the $n \times n$ matrices, i.e. the square matrices. In this case we also say the matrix is of order n. I_n denotes the $n \times n$ unit matrix (also called identity matrix). The zero matrix is denoted by 0.

Let V be a vector space of finite dimension n, over the field $\mathbf{R}$ of real numbers, or the field $\mathbf{C}$ of complex numbers. If there is no need to distinguish between the two, we will speak of the field K of *scalars*. A *basis* of V is a set $\{\, \mathbf{e}_1, \mathbf{e}_2, \ldots, \mathbf{e}_n \,\}$ of n linearly independent vectors of V, denoted by $(\mathbf{e}_i)_{i=1}^n$. Every vector $\mathbf{v} \in V$ then has the unique representation

$$\mathbf{v} = \sum_{i=1}^{n} v_i \mathbf{e}_i$$

the scalars v_i, which we will sometimes denote by $(\mathbf{v})_i$, being the *components* of the vector $\mathbf{v}$ relative to the basis $(\mathbf{e})_i$. As long as a basis is fixed unambiguously, it is thus always possible to identify V with K^n. In matrix notation, the vector $\mathbf{v}$ will always be represented by the *column vector*

$$\mathbf{v} = \begin{pmatrix} v_1 \\ v_2 \\ \vdots \\ v_n \end{pmatrix}$$

while $\mathbf{v}^T$ and $\mathbf{v}^*$ will denote the following *row vectors*:

$$\mathbf{v}^T = (v_1, v_2, \cdots, v_n), \qquad \mathbf{v}^* = (\bar{v}_1, \bar{v}_2, \cdots, \bar{v}_n)$$

where $\bar{\alpha}$ is the complex conjugate of α. The row vector $\mathbf{v}^T$ is the *transpose* of the column vector $\mathbf{v}$, and the row vector $\mathbf{v}^*$ is the *conjugate transpose* of the column vector $\mathbf{v}$.

Definition. Let $\mathbf{C}^n$ be the familiar n-dimensional vector space. Let $\mathbf{u}, \mathbf{v} \in \mathbf{C}^n$. Then the *scalar* (or *inner*) *product* is defined as

$$(\mathbf{u}, \mathbf{v}) := \sum_{j=1}^{n} u_j \bar{v}_j.$$

Obviously

$$(\mathbf{u}, \mathbf{v}) = \overline{(\mathbf{v}, \mathbf{u})}$$

and

$$(\mathbf{u}_1 + \mathbf{u}_2, \mathbf{v}) = (\mathbf{u}_1, \mathbf{v}) + (\mathbf{u}_2, \mathbf{v}).$$

In matrix notation the scalar product can be written as

$$(\mathbf{u}, \mathbf{v}) \equiv \mathbf{u}^T \bar{\mathbf{v}}.$$

Definition. Two vectors $\mathbf{u}, \mathbf{v} \in \mathbf{C}^n$ are called *orthogonal* if

$$(\mathbf{u}, \mathbf{v}) = 0.$$

Example. Let

$$\mathbf{u} = \begin{pmatrix} 1 \\ 1 \end{pmatrix}, \qquad \mathbf{v} = \begin{pmatrix} 1 \\ -1 \end{pmatrix}.$$

Then $(\mathbf{u}, \mathbf{v}) = 0$. ♣

The scalar product induces a *norm* of $\mathbf{u}$ defined by

$$\|\mathbf{u}\| := \sqrt{(\mathbf{u}, \mathbf{u})}.$$

In section 1.9 a detailed discussion of norms is given.

Definition. A vector $\mathbf{u} \in \mathbf{C}^n$ is called *normalized* if $(\mathbf{u}, \mathbf{u}) = 1$.

Example. The vectors

$$\mathbf{u} = \frac{1}{\sqrt{2}} \begin{pmatrix} 1 \\ 1 \end{pmatrix}, \qquad \mathbf{v} = \frac{1}{\sqrt{2}} \begin{pmatrix} 1 \\ -1 \end{pmatrix}$$

are normalized. ♣

Let V and W be two vector spaces over the same field, equipped with bases $(\mathbf{e}_j)_{j=1}^n$ and $(\mathbf{f}_i)_{i=1}^m$, respectively. Relative to these bases, a linear transformation

$$\mathcal{A} : V \to W$$

is represented by the matrix having m rows and n columns

$$A = \begin{pmatrix} a_{11} & a_{12} & \cdots & a_{1n} \\ a_{21} & a_{22} & \cdots & a_{2n} \\ \vdots & \vdots & & \vdots \\ a_{m1} & a_{m2} & \cdots & a_{mn} \end{pmatrix}.$$

The elements a_{ij} of the matrix A are defined uniquely by the relations

$$\mathcal{A}\mathbf{e}_j = \sum_{i=1}^m a_{ij}\mathbf{f}_i, \qquad j = 1, 2, \ldots, n.$$

Equivalently, the jth column vector

$$\begin{pmatrix} a_{1j} \\ a_{2j} \\ \vdots \\ a_{mj} \end{pmatrix}$$

of the matrix A represents the vector $\mathcal{A}e_j$ relative to the basis $(\mathbf{f}_i)_{i=1}^m$. We call

$$(a_{i1}\, a_{i2} \cdots a_{in})$$

the ith row vector of the matrix A. A matrix with m rows and n columns is called a matrix of type (m, n), and the vector space over the field K consisting of matrices of type (m, n) with elements in K is denoted by $\mathcal{A}_{m,n}$. A column vector is then a matrix of type $(m, 1)$ and a row vector a matrix of type $(1, n)$. A matrix is called real or complex according whether its elements are in the field $\mathbf{R}$ or the field $\mathbf{C}$. A matrix A with elements a_{ij} is written as

$$A = (a_{ij})$$

the first index i always designating the row and the second, j, the column.

The *null matrix* (also called the *zero matrix*) is represented by the symbol 0.

Definition. Given a matrix $A \in \mathcal{A}_{m,n}(\mathbf{C})$, the matrix $A^* \in \mathcal{A}_{n,m}(\mathbf{C})$ denotes the *adjoint* of the matrix A and is defined uniquely by the relations

$$(A\mathbf{u}, \mathbf{v})_m = (\mathbf{u}, A^*\mathbf{v})_n \quad \text{for every } \mathbf{u} \in \mathbf{C}^n, \quad \mathbf{v} \in \mathbf{C}^m$$

which imply that $(A^*)_{ij} = \bar{a}_{ji}$.

Definition. Given a matrix $A = \mathcal{A}_{m,n}(\mathbf{R})$, the matrix $A^T \in \mathcal{A}_{n,m}(\mathbf{R})$ denotes the *transpose* of a matrix A and is defined uniquely by the relations

$$(A\mathbf{u}, \mathbf{v})_m = (\mathbf{u}, A^T\mathbf{v})_n \text{ for every } \mathbf{u} \in \mathbf{R}^n, \quad \mathbf{v} \in \mathbf{R}^m$$

which imply that $(A^T)_{ij} = a_{ji}$.

To the composition of linear transformations there corresponds the multiplication of matrices.

Definition. If $A = (a_{ij})$ is a matrix of type (m, l) and $B = (b_{kj})$ of type (l, n), their *matrix product* AB is the matrix of type (m, n) defined by

$$(AB)_{ij} = \sum_{k=1}^{l} a_{ik} b_{kj}.$$

Recall that

$$(AB)^T = B^T A^T, \qquad (AB)^* = B^* A^*.$$

Note that $AB \neq BA$, in general, where A and B are $n \times n$ matrices.

Definition. A matrix of type (n, n) is said to be *square*, or a matrix of *order* n if it is desired to make explicit the integer n; it is convenient to speak of a matrix as retangular if it is not necessarily square.

Definition. If $A = (a_{ij})$ is a square matrix, the elements a_{ii} are called *diagonal elements*, and the elements $a_{ij}, i \neq j$, are called *off-diagonal elements.*

Definition. The *identity matrix* (also called *unit matrix*) is the matrix

$$I := (\delta_{ij}).$$

Definition. A matrix A is *invertible* if there exists a matrix (which is unique, if it does exist), written as A^{-1} and called the *inverse* of the matrix A, which satisfies

$$AA^{-1} = A^{-1}A = I.$$

Otherwise, the matrix is said to be *singular.*

Recall that if A and B are invertible matrices

$$(AB)^{-1} = B^{-1}A^{-1}, \qquad (A^T)^{-1} = (A^{-1})^T, \qquad (A^*)^{-1} = (A^{-1})^*.$$

Definition. A matrix A is *symmetric* if A is real and $A = A^T$.

Definition. A matrix A is *Hermitian* if $A = A^*$.

Definition. A matrix A is *orthogonal* if A is real and $AA^T = A^T A = I$.

Definition. A matrix A is *unitary* if $AA^* = A^* A = I$.

Example. Consider the matrix

$$A = \begin{pmatrix} 0 & i \\ -i & 0 \end{pmatrix}.$$

The matrix A is Hermitian and unitary. We have $A^* = A$ and $A^* = A^{-1}$. ♣

Definition. A matrix is *normal* if $AA^* = A^* A$.

Example. The matrix

$$A = \begin{pmatrix} 0 & i \\ -i & 0 \end{pmatrix}$$

is normal, whereas

$$B = \begin{pmatrix} 0 & i \\ 0 & 0 \end{pmatrix}$$

is not a normal matrix. Note that $B^* B$ is normal. Normal matrices include diagonal, real symmetric, real skew-symmetric, orthogonal, Hermitian, skew-hermitian, and unitary matrices. ♣

Definition. A matrix $A = (a_{ij})$ is *diagonal* if $a_{ij} = 0$ for $i \neq j$ and is written as

$$A = \text{diag}\ (a_{ii}) = \text{diag}\ (a_{11}, a_{22}, \ldots a_{nn}).$$

The matrix product of two $n \times n$ diagonal matrices is again a diagonal matrix.

Definition. Let $A = (a_{ij})$ be an $m \times n$ matrix over a field K. The columns of A generate a subspace of K^m, whose dimension is called the column rank of A. The rows generate a subspace of K^n whose dimension is called the row rank of A. In other words: the column rank of A is the maximum number of linearly independent columns, and the row rank is the maximum number of linearly independent rows. The row rank and the column rank of A are equal to the same number r. Thus r is simply called the rank of the matrix A.

Example. The rank of the matrix

$$A = \begin{pmatrix} 1 & 2 & 3 \\ 4 & 5 & 6 \end{pmatrix}$$

is $r(A) = 2$. The rank of the matrix product of two matrices cannot exceed the rank of either factors. ♣

Exercises. (1) Let A, B be $n \times n$ upper triangular matrices. Is $AB = BA$?

(2) Let A be an arbitrary $n \times n$ matrix. Let B be a diagonal matrix. Is $AB = BA$?

(3) Let A be a normal matrix and U be a unitary matrix. Show that U^*AU is a normal matrix.

(4) Show that the following operations, called elementary transformations, on a matrix do not change its rank:
(i) The interchange of the i-th and j-th rows.
(ii) The interchange of the i-th and j-th columns.

(5) Let A and B be two square matrices of the same order. Is it possible to have $AB + BA = 0$?

(6) Let A_k, $1 \leq k \leq m$, be matrices of order n satisfying

$$\sum_{k=1}^{m} A_k = I.$$

Show that the following conditions are equivalent.

(i) $A_k = (A_k)^2, \quad 1 \leq k \leq m$

(ii) $A_k A_l = 0$ for $k \neq l, \quad 1 \leq k, l \leq m$

(iii) $\sum_{k=1}^{m} r(A_k) = n$

denoting by $r(B)$ the rank of the matrix B.

(7) Prove that if A is of order $m \times n$, B is of order $n \times p$ and C is of order $p \times q$, then

$$A(BC) = (AB)C.$$

1.2 Trace and Determinant

In this section we introduce the trace and determinant of a $n \times n$ matrix and summarize their properties.

Definition. The *trace* of a square matrix $A = (a_{jk})$ of order n is defined as the sum of its diagonal elements

$$\mathrm{tr}A := \sum_{j=1}^{n} a_{jj}.$$

Example. Let

$$A = \begin{pmatrix} 1 & 2 \\ 0 & -1 \end{pmatrix}.$$

Then $\mathrm{tr}A = 0$. ♣

The properties of the trace are as follows. Let $a, b \in \mathbf{C}$ and let A, B and C be three $n \times n$ matrices. Then

$$\begin{aligned}
\mathrm{tr}(aA + bB) &= a\mathrm{tr}(A) + b\mathrm{tr}(B) \\
\mathrm{tr}(AB) &= \mathrm{tr}(BA) \\
\mathrm{tr}A &= \mathrm{tr}(S^{-1}AS) \qquad S \quad \text{nonsingular } n \times n \text{ matrix} \\
\mathrm{tr}(ABC) &= \mathrm{tr}(CAB) = \mathrm{tr}(BCA).
\end{aligned}$$

The last property is called the *cyclic invariance* of the trace. Notice, however, that

$$\mathrm{tr}(ABC) \neq \mathrm{tr}(BAC)$$

in general. An example is given by the following matrices

$$A = \begin{pmatrix} 1 & 0 \\ 0 & 0 \end{pmatrix}, \qquad B = \begin{pmatrix} 0 & 1 \\ 0 & 0 \end{pmatrix}, \qquad C = \begin{pmatrix} 0 & 0 \\ 1 & 0 \end{pmatrix}.$$

We have $\mathrm{tr}(ABC) = 1$ but $\mathrm{tr}(BAC) = 0$.

If λ_j, $j = 1, 2, \ldots, n$ are the eigenvalues of A (see section 1.3), then

$$\mathrm{tr}A = \sum_{j=1}^{n} \lambda_j.$$

More generally, if p designates a polynomial

$$p(x) = \sum_{j=0}^{r} a_j x^j$$

then

$$\operatorname{tr}(p(A)) = \sum_{k=1}^{n} p(\lambda_k).$$

Moreover we find

$$\operatorname{tr}(AA^*) = \operatorname{tr}(A^*A) = \sum_{j,k=1}^{n} |a_{jk}|^2 \geq 0.$$

Thus $\operatorname{tr}(AA^*)$ can be viewed as a norm of A.

Next we introduce the definition of the determinant of an $n \times n$ matrix. Then we give the properties of the determinant.

Definition. The *determinant* of an $n \times n$ matrix A is a scalar quantity denoted by $\det A$ and is given by

$$\det A := \sum_{j_1, j_2, \cdots, j_n} p(j_1, j_2, \cdots j_n) a_{1j_1} a_{2j_2} \cdots a_{nj_n}$$

where $p(j_1, j_2, \cdots, j_n)$ is a permutation equal to ± 1 and the summation extends over $n!$ permutations $j_1, j_2, \ldots, j_n$ of the integers $1, 2, \ldots, n$. For an $n \times n$ matrix there exist $n!$ permutations. Therefore

$$p(j_1, j_2, \cdots, j_n) = \operatorname{sign} \prod_{1 \leq s < r \leq n} (j_r - j_s).$$

Example. For a matrix of order (3,3) we find

$$\begin{aligned} &p(1,2,3) = 1, \quad &p(1,3,2) = -1, \quad &p(3,1,2) = 1 \\ &p(3,2,1) = -1, \quad &p(2,3,1) = 1, \quad &p(2,1,3) = -1. \end{aligned}$$

Then the determinant for a 3×3 matrix is given by

$$\det \begin{pmatrix} a_{11} & a_{12} & a_{13} \\ a_{21} & a_{22} & a_{23} \\ a_{31} & a_{32} & a_{33} \end{pmatrix}$$

$$= a_{11}a_{22}a_{33} - a_{11}a_{23}a_{32} + a_{13}a_{21}a_{32} - a_{13}a_{22}a_{31} + a_{12}a_{23}a_{31} - a_{12}a_{21}a_{33}. \quad \clubsuit$$

Definition. We call a square matrix A a *nonsingular* matrix if

$$\det A \neq 0$$

whereas if $\det A = 0$ the matrix A is called a *singular* matrix.

If $\det A \neq 0$, then A^{-1} exists. Conversly, if A^{-1} exists, then $\det A \neq 0$.

Example. The matrix

$$\begin{pmatrix} 1 & 1 \\ 1 & 0 \end{pmatrix}$$

is nonsingular and the matrix

$$\begin{pmatrix} 0 & 1 \\ 0 & 0 \end{pmatrix}$$

is singular. ♣

Next we list some properties of determinants.

1. Let A be an $n \times n$ matrix and A^T the transpose. Then

$$\det A = \det A^T.$$

Example.

$$\det \begin{pmatrix} 1 & 0 & -1 \\ 2 & 1 & 0 \\ 1 & -1 & 2 \end{pmatrix} = \det \begin{pmatrix} 1 & 2 & 1 \\ 0 & 1 & -1 \\ -1 & 0 & 2 \end{pmatrix} = 5. \quad ♣$$

Remark. Let

$$A = \begin{pmatrix} 0 & -i \\ 1 & 0 \end{pmatrix}.$$

Then

$$A^T = \begin{pmatrix} 0 & 1 \\ -i & 0 \end{pmatrix}$$

and

$$A^* \equiv \bar{A}^T = \begin{pmatrix} 0 & 1 \\ i & 0 \end{pmatrix}.$$

Obviously

$$\det A \neq \det A^*.$$

2. Let A be an $n \times n$ matrix and $\alpha \in \mathbf{R}$. Then

$$\det(\alpha A) = \alpha^n \det A.$$

3. Let A be an $n \times n$ matrix. If two adjacent columns are equal, i.e. $A_j = A_{j+1}$ for some $j = 1, \ldots n-1$, then $\det A = 0$.

4. If any vector in A is a zero vector then $\det A = 0$.

5. Let A be an $n \times n$ matrix. Let j be some integer, $1 \leq j < n$. If the j-th and $(j+1)$-th columns are interchanged, then the determinant changes by a sign.

6. Let $A_1, \ldots, A_n$ be the column vectors of an $n \times n$ matrix. If they are linearly dependent, then $\det A = 0$.

7. Let A and B be $n \times n$ matrices. Then

$$\det(AB) = \det(A)\det(B).$$

8. Let A be an $n \times n$ diagonal matrix. Then

$$\det A = a_{11}a_{22}\cdots a_{nn}.$$

9. $(d/dt)\det(A(t))$ = sum of the determinants where each of them is obtained by differentiating the rows of A with respect to t one at a time, then taking its determinant.

Proof. Since

$$\det A(t) = \sum_{j_1,\cdots,j_n} p(j_1,\cdots,j_n)a_{1j_1}(t)\cdots a_{nj_n}(t)$$

we find

$$\begin{aligned}
\frac{d}{dt}\det(A(t)) &= \sum_{j_1,\cdots,j_n} p(j_1,\cdots,j_n)\frac{da_{1j_1}(t)}{dt}a_{2j_2}(t)\cdots a_{nj_n}(t) + \\
&+ \sum_{j_1,\cdots,j_n} p(j_1,\cdots,j_n)a_{1j_1}(t)\frac{da_{2j_2}(t)}{dt}\cdots a_{nj_n}(t) + \cdots \\
&+ \sum_{j_1,\cdots,j_n} p(j_1,\cdots,j_n)a_{1j_1}(t)\cdots a_{n-1j_{n-1}}(t)\frac{da_{nj_n}(t)}{dt}. \quad \spadesuit
\end{aligned}$$

Example.

$$\frac{d}{dt}\det\begin{pmatrix} e^t & \cos t \\ 1 & \sin t^2 \end{pmatrix} = \det\begin{pmatrix} e^t & -\sin t \\ 1 & \sin t^2 \end{pmatrix} + \det\begin{pmatrix} e^t & \cos t \\ 0 & 2t\cos t^2 \end{pmatrix}. \quad \clubsuit$$

10. Let A be an invertible $n \times n$ symmetric matrix. Then

$$\mathbf{v}^T A^{-1}\mathbf{v} = \frac{\det(A + \mathbf{v}\mathbf{v}^T)}{\det A} - 1$$

for every vector $\mathbf{v} \in \mathbf{R}^n$.

Example. Let

$$A = \begin{pmatrix} 0 & 1 \\ 1 & 0 \end{pmatrix}, \qquad \mathbf{v} = \begin{pmatrix} 1 \\ 1 \end{pmatrix}.$$

Then $A^{-1} = A$ and therefore

$$\mathbf{v}^T A^{-1} \mathbf{v} = 2.$$

Since

$$\mathbf{v}\mathbf{v}^T = \begin{pmatrix} 1 & 1 \\ 1 & 1 \end{pmatrix}$$

and $\det A = -1$ we obtain

$$\frac{\det(A + \mathbf{v}\mathbf{v}^T)}{\det A} - 1 = 2. \quad \clubsuit$$

11. Let A be an $n \times n$ matrix. Then

$$\det(\exp A) \equiv \exp(\mathrm{tr} A).$$

12. Let A be an $n \times n$ matrix. Let $\lambda_1, \lambda_2, \ldots, \lambda_n$ be the eigenvalues of A (see section 1.3). Then

$$\det A = \lambda_1 \lambda_2 \cdots \lambda_n.$$

13. Let A be a Hermitian matrix. Then $\det(A)$ is a real number.

14. Let A, B, C be $n \times n$ matrices. Then

$$\det \begin{pmatrix} A & 0 \\ C & B \end{pmatrix} = \det A \det B.$$

15. The determinant of the matrix

$$A_n := \begin{pmatrix} b_1 & a_2 & 0 & \ldots & 0 & 0 \\ -1 & b_2 & a_3 & \ldots & 0 & 0 \\ \vdots & \vdots & \vdots & \ddots & \vdots & \vdots \\ 0 & 0 & 0 & \ldots & b_{n-1} & a_n \\ 0 & 0 & 0 & \ldots & -1 & b_n \end{pmatrix}, \qquad n = 1, 2, \ldots$$

satisfies the recursion relation

$$\det(A_n) = b_n \det(A_{n-1}) + a_n \det(A_{n-2}), \quad \det(A_0) = 1, \quad \det(A_1) = b_1, \qquad n = 2, 3, \ldots$$

Exercises. (1) Let X and Y be $n \times n$ matrices over $\mathbf{R}$. Show that

$$(X, Y) := \operatorname{tr}(XY^T)$$

defines a scalar product.

(2) Let A and B be $n \times n$ matrices. Show that

$$\operatorname{tr}([A, B]) = 0$$

where $[\, ,\,]$ denotes the commutator (i.e. $[A, B] := AB - BA$).

(3) Use (2) to show that the relation

$$[A, B] = \lambda I, \qquad \lambda \in \mathbf{C}$$

for finite-dimensional matrices can only be satisfied if $\lambda = 0$. For certain infinite dimensional matrices A and B we can find a nonzero λ.

(4) Let A and B be $n \times n$ matrices. Suppose that AB is nonsingular. Show that A and B are nonsingular matrices.

(5) Let A and B be $n \times n$ matrices over $\mathbf{R}$. Assume that A is skew-symmetric, i.e. $A^T = -A$. Assume that n is odd. Show that $\det(A) = 0$.

(6) Let

$$A = \begin{pmatrix} A_{11} & A_{12} \\ A_{21} & A_{22} \end{pmatrix}$$

be a square matrix partitioned into blocks. Assuming the submatrix A_{11} to be invertible, show that

$$\det A = \det A_{11} \det(A_{22} - A_{21} A_{11}^{-1} A_{12}).$$

(7) A square matrix A for which $A^n = 0$, where n is a positive integer, is called *nilpotent*. Let A be a nilpotent matrix. Show that $\det A = 0$.

1.3 Eigenvalue Problem

The eigenvalue problem plays a central role in theoretical physics. We give a short introduction into the eigenvalue calculation for finite dimensional matrices. Then we study the eigenvalue problem for Kronecker products of matrices.

Definition. A complex number λ is said to be an *eigenvalue* (or *characteristic value*) of an $n \times n$ matrix A, if there is at least one nonzero vector $\mathbf{u} \in \mathbf{C}^n$ satisfying the *eigenvalue equation*

$$A\mathbf{u} = \lambda\mathbf{u}.$$

Each nonzero vector $\mathbf{u} \in \mathbf{C}^n$ satisfying the eigenvalue equation is called an *eigenvector* (or *characteristic vector*) of A with eigenvalue λ.

The eigenvalue equation can be written as

$$(A - \lambda I)\mathbf{u} = \mathbf{0}$$

where I is the $n \times n$ unit matrix and $\mathbf{0}$ is the zero vector.

This system of n linear simultaneous equations in $\mathbf{u}$ has a non-trivial solution for the vector $\mathbf{u}$ only if the matrix $(A - \lambda I)$ is singular, i.e.

$$\det(A - \lambda I) = 0.$$

The expansion of the determinant gives a polynomial in λ of degree equal to n, which is called the characteristic polynomial of the matrix A. The n roots of the equation $\det(A - \lambda I) = 0$, called the *characteristic equation*, are the eigenvalues of A.

Definition. The *spectrum* of the matrix A is the subset

$$\mathrm{sp}(A) := \bigcup_{i=1}^{n} \{\, \lambda_i(A) \,\}$$

of the complex plane. The *spectral radius* of the matrix A is the non-negative number defined by

$$\varrho(A) := \max\{\, |\lambda_i(A)| \;:\; 1 \le i \le n \,\}.$$

If $\lambda \in \mathrm{sp}(A)$, the vector subspace

$$\{\, \mathbf{v} \in V \;:\; A\mathbf{v} = \lambda\mathbf{v} \,\}$$

(of dimension at least 1) is called the *eigenspace* corresponding to the eigenvalue λ.

Example. Let

$$A = \begin{pmatrix} 0 & -i \\ i & 0 \end{pmatrix}.$$

Then

$$\det(A - \lambda I) \equiv \lambda^2 - 1 = 0.$$

Therefore the eigenvalues are given by $\lambda_1 = 1$, $\lambda_2 = -1$. To find the eigenvector of the eigenvalue $\lambda_1 = 1$ we have to solve

$$\begin{pmatrix} 0 & -i \\ i & 0 \end{pmatrix} \begin{pmatrix} u_1 \\ u_2 \end{pmatrix} = 1 \begin{pmatrix} u_1 \\ u_2 \end{pmatrix}.$$

Therefore

$$-iu_2 = u_1, \qquad iu_1 = u_2$$

and the eigenvector of $\lambda_1 = 1$ is given by

$$\mathbf{u}_1 = \begin{pmatrix} 1 \\ i \end{pmatrix}.$$

For $\lambda_2 = -1$ we have

$$\begin{pmatrix} 0 & -i \\ i & 0 \end{pmatrix} \begin{pmatrix} u_1 \\ u_2 \end{pmatrix} = -1 \begin{pmatrix} u_1 \\ u_2 \end{pmatrix}$$

and hence

$$\mathbf{u}_2 = \begin{pmatrix} 1 \\ -i \end{pmatrix}.$$

We see that $(\mathbf{u}_1, \mathbf{u}_2) = 0$. ♣

A special role in theoretical physics is played by the Hermitian matrices. In this case we have the following theorem.

Theorem. Let A be a Hermitian matrix, i.e. $A^* = A$, where $A^* \equiv \bar{A}^T$. The eigenvalues of A are real, and two eigenvectors corresponding to two different eigenvalues are mutually orthogonal.

Proof. The eigenvalue equation is $A\mathbf{u} = \lambda\mathbf{u}$, where $\mathbf{u} \neq \mathbf{0}$. Now we have the identity

$$(A\mathbf{u})^*\mathbf{u} \equiv \mathbf{u}^*A^*\mathbf{u} \equiv \mathbf{u}^*(A^*\mathbf{u}) \equiv \mathbf{u}^*(A\mathbf{u})$$

since A is Hermitian, i.e. $A = A^*$. Inserting the eigenvalue equation into this equation yields

$$(\lambda\mathbf{u})^*\mathbf{u} = \mathbf{u}^*(\lambda\mathbf{u})$$

or

$$\bar{\lambda}(\mathbf{u}^*\mathbf{u}) = \lambda(\mathbf{u}^*\mathbf{u}).$$

Since $\mathbf{u}^*\mathbf{u} \neq 0$, we have $\bar{\lambda} = \lambda$ and therefore λ must be real. Let

$$A\mathbf{u}_1 = \lambda_1\mathbf{u}_1, \qquad A\mathbf{u}_2 = \lambda_2\mathbf{u}_2.$$

Now

$$\lambda_1(\mathbf{u}_1, \mathbf{u}_2) = (\lambda_1 \mathbf{u}_1, \mathbf{u}_2) = (A\mathbf{u}_1, \mathbf{u}_2) = (\mathbf{u}_1, A\mathbf{u}_2) = (\mathbf{u}_1, \lambda_2 \mathbf{u}_2) = \lambda_2(\mathbf{u}_1, \mathbf{u}_2).$$

Since $\lambda_1 \neq \lambda_2$, we find that $(\mathbf{u}_1, \mathbf{u}_2) = 0$. ♠

Theorem. The eigenvalues λ_j of a unitary matrix satisfy $|\lambda_j| = 1$.

Proof. Since U is a unitary matrix we have

$$U^* = U^{-1}$$

where U^{-1} is the inverse of U. Let

$$U\mathbf{u} = \lambda \mathbf{u}$$

be the eigenvalue equation. It follows that

$$(U\mathbf{u})^* = (\lambda \mathbf{u})^*$$

or

$$\mathbf{u}^* U^* = \bar{\lambda} \mathbf{u}^*.$$

Thus we obtain

$$\mathbf{u}^* U^* U \mathbf{u} = \bar{\lambda}\lambda \mathbf{u}^* \mathbf{u}.$$

Since $U^*U = I$ we obtain

$$\mathbf{u}^* \mathbf{u} = \bar{\lambda}\lambda \mathbf{u}^* \mathbf{u}.$$

Since $\mathbf{u}^* \mathbf{u} \neq 0$ we have

$$\bar{\lambda}\lambda = 1.$$

Thus λ can be written as

$$\lambda = \exp(i\alpha), \qquad \alpha \in \mathbf{R}.$$

Thus $|\lambda| = 1$. ♠

Exercises. (1) Show that the eigenvectors corresponding to distinct eigenvalues are linearly independent.

(2) Show that

$$\operatorname{tr}(A) = \sum_{i=1}^{n} \lambda_i(A), \qquad \det(A) = \prod_{i=1}^{n} \lambda_i(A).$$

(3) Show that an $n \times n$ matrix satisfies its own characteristic equation. This is the so-called *Cayley-Hamilton theorem.*

(4) Let A be an invertible matrix whose elements, as well as those of A^{-1}, are all non-negative. Show that there exists a permutation matrix P and a matrix $D = \operatorname{diag}(d_j)$, with d_j positive, such that $A = PD$ (the converse is obvious).

(5) Let A and B be two square matrices of the same order. Show that the matrices AB and BA have the same characteristic polynomial.

(6) Let $a, b, c \in \mathbf{R}$. Find the eigenvalues and eigenvectors of the matrix

$$A = \begin{pmatrix} a & b & & & & \\ c & a & b & & & \\ & c & a & b & & \\ & & \ddots & \ddots & \ddots & \\ & & & c & a & b \\ & & & & c & a \end{pmatrix}.$$

(7) Let $a_1, a_2, \ldots, a_n \in \mathbf{R}$. Show that the eigenvalues of the matrix

$$A = \begin{pmatrix} a_1 & a_2 & \cdots & a_{n-1} & a_n \\ a_n & a_1 & a_2 & \cdots & a_{n-1} \\ a_{n-1} & a_n & a_1 & & \\ \vdots & \cdots & \cdots & \cdots & \vdots \\ a_3 & \cdots & a_n & a_1 & a_2 \\ a_2 & a_3 & \cdots & a_n & a_1 \end{pmatrix}$$

called a *circulant matrix*, are of the form

$$\lambda_{l+1} = a_1 + a_2\xi_l + a_3\xi_l^2 + \cdots + a_n\xi_l^{n-1}, \qquad l = 0, 1, \ldots, n-1$$

where $\xi_l := e^{2i\pi l/n}$.

1.4 Projection Matrices

First we introduce the definition of a projection matrix and give some of its properties. Projection matrices (projection operators) play a central role in finite group theory in the decomposition of Hilbert spaces into invariant subspaces (Steeb [38]).

Definition. An $n \times n$ matrix Π is called a *projection matrix* if

$$\Pi = \Pi^*$$

and

$$\Pi^2 = \Pi.$$

The element $\Pi\mathbf{u}$ ($\mathbf{u} \in \mathbf{C}^n$) is called the *projection* of the element $\mathbf{u}$.

Example. Let $n = 2$ and

$$\Pi_1 = \begin{pmatrix} 1 & 0 \\ 0 & 0 \end{pmatrix}, \qquad \Pi_2 = \begin{pmatrix} 0 & 0 \\ 0 & 1 \end{pmatrix}.$$

Then $\Pi_1^* = \Pi_1$, $\Pi_1^2 = \Pi_1$, $\Pi_2^* = \Pi_2$ and $\Pi_2^2 = \Pi_2$. Furthermore $\Pi_1\Pi_2 = 0$ and

$$\Pi_1 \begin{pmatrix} u_1 \\ u_2 \end{pmatrix} = \begin{pmatrix} u_1 \\ 0 \end{pmatrix}, \qquad \Pi_2 \begin{pmatrix} u_1 \\ u_2 \end{pmatrix} = \begin{pmatrix} 0 \\ u_2 \end{pmatrix}. \quad \clubsuit$$

Theorem. Let Π_1 and Π_2 be two $n \times n$ projection matrices. Assume that $\Pi_1\Pi_2 = 0$. Then

$$(\Pi_1\mathbf{u}, \Pi_2\mathbf{u}) = 0.$$

Proof. We find

$$(\Pi_1\mathbf{u}, \Pi_2\mathbf{u}) = (\Pi_1\mathbf{u})^*(\Pi_1\mathbf{u}) = (\mathbf{u}^*\Pi_1^*)(\Pi_2\mathbf{u}) = \mathbf{u}^*(\Pi_1\Pi_2)\mathbf{u} = 0. \quad \spadesuit$$

Theorem. Let I be the $n \times n$ unit matrix and Π be a projection matrix. Then $I - \Pi$ is a projection matrix.

Proof. Since

$$(I - \Pi)^* = I^* - \Pi^* = I - \Pi$$

and

$$(I - \Pi)^2 = (I - \Pi)(I - \Pi) = I - \Pi - \Pi + \Pi = I - \Pi$$

we find that $I - \Pi$ is a projection matrix. $\spadesuit$

Theorem. The eigenvalues λ_j of a projection matrix Π are given by $\lambda_j \in \{0, 1\}$.

Proof. From the eigenvalue equation

$$\Pi \mathbf{u} = \lambda \mathbf{u}$$

we find

$$\Pi(\Pi \mathbf{u}) = (\Pi\Pi)\mathbf{u} = \lambda \Pi \mathbf{u}.$$

Using the fact that $\Pi^2 = \Pi$ we obtain

$$\Pi \mathbf{u} = \lambda^2 \mathbf{u}.$$

Thus

$$\lambda = \lambda^2$$

since $\mathbf{u} \neq \mathbf{0}$. Thus $\lambda \in \{0, 1\}$.

Projection Theorem. Let U be a non-empty, convex, closed subset of the vector space $\mathbf{C}^n$. Given any element $\mathbf{w} \in \mathbf{C}^n$, there exists a unique element $\Pi\mathbf{w}$ such that

$$\Pi\mathbf{w} \in U \quad \text{and} \quad ||\mathbf{w} - \Pi\mathbf{w}|| = \inf_{\mathbf{v} \in U} ||\mathbf{w} - \mathbf{v}||.$$

This element $\Pi\mathbf{w} \in U$ satisfies

$$(\Pi\mathbf{w} - \mathbf{w}, \mathbf{v} - \Pi\mathbf{w}) \geq 0 \quad \text{for every } \mathbf{v} \in U$$

and, conversely, if any element $\mathbf{u}$ satisfies

$$\mathbf{u} \in U \quad \text{and} \quad (\mathbf{u}, \mathbf{v} - \mathbf{u}) \geq 0 \quad \text{for every } \mathbf{v} \in U$$

then

$$\mathbf{u} = \Pi\mathbf{w}.$$

Furthermore

$$||\Pi\mathbf{u} - \Pi\mathbf{v}|| \leq ||\mathbf{u} - \mathbf{v}||.$$

For the proof refer to Ciarlet [9].

Exercises. (1) Show that the matrices

$$\Pi_1 = \frac{1}{2}\begin{pmatrix} 1 & 1 \\ 1 & 1 \end{pmatrix}, \qquad \Pi_2 = \frac{1}{2}\begin{pmatrix} 1 & -1 \\ -1 & 1 \end{pmatrix}$$

are projection matrices and that $\Pi_1 \Pi_2 = 0$.

(2) Is the sum of two $n \times n$ projection matrices an $n \times n$ projection matrix ?

(3) Let A be an $n \times n$ matrix with $A^2 = A$. Show that $\det A$ is either equal to zero or equal to 1.

(4) Let

$$\Pi = \frac{1}{4}\begin{pmatrix} 1 & 1 & 1 & 1 \\ 1 & 1 & 1 & 1 \\ 1 & 1 & 1 & 1 \\ 1 & 1 & 1 & 1 \end{pmatrix}$$

and

$$\mathbf{u} = \begin{pmatrix} 1 \\ 0 \\ 0 \\ 0 \end{pmatrix}, \qquad \mathbf{v} = \begin{pmatrix} 0 \\ 0 \\ 0 \\ 1 \end{pmatrix}.$$

Show that

$$||\Pi\mathbf{u} - \Pi\mathbf{v}|| \leq ||\mathbf{u} - \mathbf{v}||.$$

(5) Consider the matrices

$$A = \begin{pmatrix} 2 & 1 \\ 1 & 2 \end{pmatrix}, \qquad I = \begin{pmatrix} 1 & 0 \\ 0 & 1 \end{pmatrix}, \qquad C = \begin{pmatrix} 0 & 1 \\ 1 & 0 \end{pmatrix}.$$

Show that $[A, I] = 0$ and $[A, C] = 0$. Show that $IC = C$, $CI = C$, $CC = I$. A group theoretical reduction (Steeb [38]) leads to the projection matrices

$$\Pi_1 = \frac{1}{2}\begin{pmatrix} 1 & 1 \\ 1 & 1 \end{pmatrix}, \qquad \Pi_2 = \frac{1}{2}\begin{pmatrix} 1 & -1 \\ -1 & 1 \end{pmatrix}.$$

Apply the projection operators to the standard basis to find a new basis. Show that the matrix A takes the form

$$\widetilde{A} = \begin{pmatrix} 3 & 0 \\ 0 & 1 \end{pmatrix}$$

within the new basis. Notice that the new basis must be normalized before the matrix $\widetilde{A}$ can be calculated.

1.5 Fourier and Hadamard Matrices

Fourier and Hadamard matrices play an important role in spectral analysis (Davis [10], Elliott and Rao [13], Regalia and Mitra [29]). We give a short introduction to these types of matrices. In section 2.9 we discuss the connection with the Kronecker product.

Let n be a fixed integer ≥ 1. We define

$$w := \exp\left(\frac{2\pi i}{n}\right) \equiv \cos\left(\frac{2\pi}{n}\right) + i\sin\left(\frac{2\pi}{n}\right)$$

where $i = \sqrt{-1}$. w might be taken as any primitive n-th root of unity. It can easily be proved that

$$\begin{aligned} w^n &= 1 \\ w\bar{w} &= 1 \\ \bar{w} &= w^{-1} \\ \bar{w}^k &= w^{-k} = w^{n-k} \end{aligned}$$

and

$$1 + w + w^2 + \cdots + w^{n-1} = 0$$

where $\bar{w}$ is the complex conjugate of w.

Definition. By the *Fourier matrix* of order n, we mean the matrix $F(= F_n)$ where

$$F^* := \frac{1}{\sqrt{n}}(w^{(i-1)(j-1)}) \equiv \frac{1}{\sqrt{n}} \begin{pmatrix} 1 & 1 & 1 & \cdots & 1 \\ 1 & w & w^2 & \cdots & w^{n-1} \\ 1 & w^2 & w^4 & \cdots & w^{2(n-1)} \\ \vdots & \vdots & \vdots & & \vdots \\ 1 & w^{n-1} & w^{2(n-1)} & \cdots & w^{(n-1)(n-1)} \end{pmatrix}$$

where F^* is the conjugate transpose of F.

The sequence w^k, $k = 0, 1, 2 \cdots$, is periodic with period n. Consequently there are only n distinct elements in F. Therefore F^* can be written as

$$F^* = \frac{1}{\sqrt{n}} \begin{pmatrix} 1 & 1 & 1 & \cdots & 1 \\ 1 & w & w^2 & \cdots & w^{n-1} \\ 1 & w^2 & w^4 & \cdots & w^{n-2} \\ \vdots & \vdots & \vdots & & \vdots \\ 1 & w^{n-1} & w^{n-2} & \cdots & w \end{pmatrix}.$$

The following theorem can easily be proved

Theorem. F is unitary, i.e.

$$FF^* = F^*F = I \Longleftrightarrow F^{-1} = F^* .$$

Proof. This is a result of the geometric series identity

$$\sum_{r=0}^{n-1} w^{r(j-k)} \equiv \frac{1 - w^{n(j-k)}}{1 - w^{j-k}} = \begin{cases} n \text{ if } & j = k \\ 0 \text{ if } & j \neq k \end{cases} . \quad \spadesuit$$

A second application of the geometrical identity yields

$$F^{*2} \equiv F^*F^* \equiv \begin{pmatrix} 1 & 0 & \cdots & & 0 \\ 0 & 0 & \cdots & & 1 \\ 0 & 0 & \cdots & 1 & 0 \\ \vdots & \vdots & & \vdots & \vdots \\ 0 & 1 & \cdots & & 0 \end{pmatrix} = F^2.$$

This means F^{*2} is an $n \times n$ permutation matrix.

Corollary.

$$F^{*4} = I, \qquad F^{*3} = F^{*4}(F^*)^{-1} = IF = F.$$

Corollary. The eigenvalues of F are ± 1, $\pm i$, with appropriate multiplicities.

The characteristic polynomials $f(\lambda)$ of $F^*(= F_n^*)$ are as follows

$$\begin{array}{lll} n \equiv 0 \text{ modulo } 4, & f(\lambda) & = & (\lambda - 1)^2(\lambda - i)(\lambda + 1)(\lambda^4 - 1)^{(n/4)-1} \\ n \equiv 1 \text{ modulo } 4, & f(\lambda) & = & (\lambda - 1)(\lambda^4 - 1)^{(1/4)(n-1)} \\ n \equiv 2 \text{ modulo } 4, & f(\lambda) & = & (\lambda^2 - 1)(\lambda^4 - 1)^{(n/4)(n-2)} \\ n \equiv 3 \text{ modulo } 4, & f(\lambda) & = & (\lambda - i)(\lambda^2 - 1)(\lambda^4 - 1)^{(1/4)(n-3)}. \end{array}$$

Definition. Let

$$Z = (z_1, z_2, \cdots, z_n)^T$$

and

$$\hat{Z} = (\hat{z}_1, \hat{z}_2, \cdots, \hat{z}_n)^T$$

where $z_j \in \mathbf{C}$. The linear transformation

$$\hat{Z} = FZ$$

where F is the Fourier matrix is called the discrete *Fourier transform.*

Its inverse transformation exists since F^{-1} exists and is given by

$$Z = F^{-1}\hat{Z} \equiv F^*\hat{Z}.$$

Let

$$p(z) = a_0 + a_1 z + \cdots + a_{n-1} z^{n-1}$$

be a polynomial of degree $\leq n-1$. It will be determined uniquely by specifying its values $p(z_n)$ at n distinct points z_k, $k = 1, 2, \cdots, n$ in the complex plane $\mathbf{C}$. Select these points z_k as the n roots of unity $1, w, w^2, \cdots, w^{n-1}$. Then

$$\sqrt{n}F^* \begin{pmatrix} a_0 \\ a_1 \\ \vdots \\ a_{n-1} \end{pmatrix} = \begin{pmatrix} p(1) \\ p(w) \\ \vdots \\ p(w^{n-1}) \end{pmatrix}$$

so that

$$\begin{pmatrix} a_0 \\ a_1 \\ \vdots \\ a_{n-1} \end{pmatrix} = \frac{1}{\sqrt{n}} F \begin{pmatrix} p(1) \\ p(w) \\ \vdots \\ p(w^{n-1}) \end{pmatrix}.$$

These formulas for interpolation at the roots of unity can be given another form.

Definition. By a *Vandermonde matrix* $V(z_0, z_1, \cdots, z_{n-1})$ is meant a matrix of the form

$$V(z_0, z_1, \ldots, z_{n-1}) := \begin{pmatrix} 1 & 1 & \cdots & 1 \\ z_0 & z_1 & \cdots & z_{n-1} \\ z_0^2 & z_1^2 & \cdots & z_{n-1}^2 \\ \vdots & \vdots & & \vdots \\ z_0^{n-1} & z_1^{n-1} & \cdots & z_{n-1}^{n-1} \end{pmatrix}.$$

It follows that

$$\begin{aligned} V(1, w, w^2, \cdots, w^{n-1}) &= n^{1/2} F^* \\ V(1, \bar{w}, \bar{w}^2, \cdots, \bar{w}^{n-1}) &= n^{1/2} \bar{F}^* = n^{1/2} F. \end{aligned}$$

Furthermore

$$\begin{aligned} p(z) &= (1, z, \cdots, z^{n-1})(a_0, a_1, \cdots, a_{n-1})^T \\ &= (1, z, \cdots, z^{n-1}) n^{-1/2} F(p(1), p(w), \cdots, p(w^{n-1}))^T \\ &= n^{-1/2}(1, z, \cdots, z^{n-1}) V(1, \bar{w}, \bar{w}^2, \cdots, \bar{w}^{n-1})(p(1), p(w), \cdots, p(w^{n-1}))^T. \end{aligned}$$

Let F'_{2^n} denote the Fourier matrices of order 2^n whose rows have been permuted according to the bit reversing permutation.

Definition. A sequence in natural order can be arranged in *bit-reversed order* as follows: For an integer expressed in binary notation, reverse the binary form and transform to decimal notation, which is then called bit-reversed notation.

Example. The number 6 can be written as

$$6 = 1 \cdot 2^2 + 1 \cdot 2^1 + 0 \cdot 2^0.$$

Therefore $6 \to 110$. Reversing the binary digits yields 011. Since

$$3 = 0 \cdot 2^2 + 1 \cdot 2^1 + 1 \cdot 2^0$$

we have $6 \to 3$. ♣

Since the sequence 0, 1 is the bit reversed order of 0, 1 and 0, 2, 1, 3 is the bit reversed order of 0, 1, 2, 3 we find that the matrices F'_2 and F'_4 are given by

$$\begin{aligned} F'_2 &= \frac{1}{\sqrt{2}} \begin{pmatrix} 1 & 1 \\ 1 & -1 \end{pmatrix} = F_2 \\ F'_4 &= \frac{1}{\sqrt{4}} \begin{pmatrix} 1 & 1 & 1 & 1 \\ 1 & -1 & 1 & -1 \\ 1 & i & -1 & -i \\ 1 & -i & -1 & i \end{pmatrix}. \end{aligned}$$

Definition. By a *Hadamard matrix* of order n, H $(\equiv H_n)$, is meant a matrix whose elements are either $+1$ or -1 and for which

$$HH^T = H^T H = nI$$

where I is the $n \times n$ unit matrix. Thus, $n^{-1/2}H$ is an orthogonal matrix.

Examples.

$$H_1 = (1)$$

$$H_2 = \sqrt{2}F_2 = \begin{pmatrix} 1 & 1 \\ 1 & -1 \end{pmatrix}$$

$$H_{4,1} = \begin{pmatrix} 1 & 1 & 1 & 1 \\ -1 & -1 & 1 & 1 \\ -1 & 1 & 1 & -1 \\ 1 & -1 & 1 & -1 \end{pmatrix}$$

$$H_{4,2} = \begin{pmatrix} 1 & 1 & 1 & -1 \\ 1 & 1 & -1 & 1 \\ 1 & -1 & 1 & 1 \\ -1 & 1 & 1 & 1 \end{pmatrix}. \quad \clubsuit$$

Sometimes the term Hadamard matrix is limited to the matrices of order 2^n. These matrices have the property

$$H_{2^n} = H_{2^n}^T$$

so that

$$H_{2^n}^2 = 2^n I.$$

A recursion relation to find H_{2^n} using the Kronecker product will be given in section 2.9.

Definition. The *Walsh-Hadamard transform* is defined as

$$\hat{Z} = HZ$$

where H is an Hadamard matrix, where $Z = (z_1, z_2, \ldots, z_n)^T$.

Exercises. (1) Show that

$$F = F^T, \qquad F^* = (F^*)^T = \bar{F}, \qquad F = \bar{F}^*.$$

This means F and F^* are symmetric.

(2) Show that the sequence

$$0, 8, 4, 12, 2, 10, 6, 14, 1, 9, 5, 13, 3, 11, 7, 15$$

is the bit reversed order of

$$0, 1, 2, 3, 4, 5, 6, 7, 8, 9, 10, 11, 12, 13, 14, 15.$$

(3) Find the eigenvalues of

$$F_4^* = \frac{1}{2}\begin{pmatrix} 1 & 1 & 1 & 1 \\ 1 & \omega & \omega^2 & \omega^3 \\ 1 & \omega^2 & \omega & \omega^2 \\ 1 & \omega^3 & \omega^2 & \omega \end{pmatrix}.$$

Derive the eigenvalues of F_4.

(4) The discrete Fourier transform in one dimension can also be written as

$$\hat{x}(k) = \frac{1}{N}\sum_{n=0}^{N-1} x(n)\exp(-ik2\pi n/N)$$

where $N \in \mathbf{N}$ and $k = 0, 1, 2, \ldots, N-1$. Show that

$$x(n) = \sum_{k=0}^{N-1} \hat{x}(k)\exp(ik2\pi n/N).$$

Let

$$x(n) = \cos(2\pi n/N)$$

where $N = 8$ and $n = 0, 1, 2, \ldots, N-1$. Find $\hat{x}(k)$.

1.6 Transformation of Matrices

Let V be a vector space of finite dimension n and let $\mathcal{A} : V \to V$ be a linear transformation, represented by a (square) matrix $A = (a_{ij})$ relative to a basis $(\mathbf{e}_i)$. Relative to another basis $(\mathbf{f}_i)$, the same transformation is represented by the matrix

$$B = Q^{-1}AQ$$

where Q is the invertible matrix whose jth column vector consists of the components of the vector $\mathbf{f}_j$ in the basis $(\mathbf{e}_i)$. Since the same linear transformation $\mathcal{A}$ can in this way be represented by different matrices, depending on the basis that is chosen, the problem arises of finding a basis relative to which the matrix representing the transformation is as simple as possible. Equivalently, given a matrix A, that is to say, those which are of the form $Q^{-1}AQ$, with Q invertible, those which have a form that is 'as simple as possible'.

Definition. If there exists an invertible matrix Q such that the matrix $Q^{-1}AQ$ is diagonal, then the matrix A is said to be *diagonalisable.*

In this case, the diagonal elements of the matrix $Q^{-1}AQ$ are the eigenvalues $\lambda_1, \lambda_2, \ldots, \lambda_n$ of the matrix A. The jth column vector of the matrix Q consists of the components (relative to the same basis as that used for the matrix A) of a normalized eigenvector corresponding to λ_j. In other words, a matrix is diagonalisable if and only if there exists a basis of eigenvectors.

Example. The matrix

$$A = \begin{pmatrix} 0 & 1 \\ 1 & 0 \end{pmatrix}$$

is diagonalizable. The matrix

$$B = \begin{pmatrix} 0 & 1 \\ 0 & 0 \end{pmatrix}$$

cannot be diagonalized. ♣

For such matrices *Jordan's theorem* gives the simplest form among all similar matrices.

Definition. A matrix $A = (a_{ij})$ of order n is *upper triangular* if $a_{ij} = 0$ for $i > j$ and *lower triangular* if $a_{ij} = 0$ for $i < j$. If there is no need to distinguish between the two, the matrix is simply called *triangular.*

Theorem. (1) Given a square matrix A, there exists a unitary matrix U such that the matrix $U^{-1}AU$ is triangular.
(2) Given a normal matrix A, there exists a unitary matrix U such that the matrix $U^{-1}AU$ is diagonal.

(3) Given a symmetric matrix A, there exists an orthogonal matrix O such that the matrix $O^{-1}AO$ is diagonal.

For the proof refer to Ciarlet [9].

The matrices U satisfying the conditions of the statement are not unique (consider, for example, $A = I$). The diagonal elements of the triangular matrix $U^{-1}AU$ of (1), or of the diagonal matrix $U^{-1}AU$ of (2), or of the diagonal matrix of (3), are the eigenvalues of the matrix A. Consequently, they are real numbers if the matrix A is Hermitian or symmetric and complex numbers of modulus 1 if the matrix is unitary or orthogonal. It follows from (2) that every Hermitian or unitary matrix is diagonalizable by a unitary matrix. The preceding argument shows that if, O is an orthogonal matrix, there exists a unitary matrix U such that $D = U^*OU$ is diagonal (the diagonal elements of D having modulus equal to 1), but the matrix U is not, in general, real, that is to say, orthogonal.

Definition. The *singular values* of a square matrix are the positive square roots of the eigenvalues of the Hermitian matrix A^*A (or A^TA, if the matrix A is real).

They are always non-negative, since from the relation $A^*A\mathbf{u} = \lambda\mathbf{u}$, $\mathbf{u} \neq \mathbf{0}$, it follows that $(A\mathbf{u})^*A\mathbf{u} = \lambda\mathbf{u}^*\mathbf{u}$.

The singular values are all strictly positive if and only if the matrix A is invertible. In fact, we have

$$A\mathbf{u} = \mathbf{0} \Rightarrow A^*A\mathbf{u} = \mathbf{0} \Rightarrow \mathbf{u}^*A^*A\mathbf{u} = (A\mathbf{u})^*A\mathbf{u} = \mathbf{0} \Rightarrow A\mathbf{u} = \mathbf{0}.$$

Definition. Two matrices A and B of type (m,n) are said to be *equivalent* if there exists an invertible matrix Q of order m and an invertible matrix R of order n such that

$$B = QAR.$$

This is a more general notion than that of the similarity of matrices. In fact, it can be shown that every square matrix is equivalent to a diagonal matrix.

Theorem. If A is a real, square matrix, there exist two orthogonal matrices U and V such that

$$U^TAV = \operatorname{diag}(\mu_i)$$

and, if A is a complex, square matrix, there exist two unitary matrices U and V such that

$$U^*AV = \operatorname{diag}(\mu_i).$$

In either case, the numbers $\mu_i \geq 0$ are the singular values of the matrix A. For the proof refer to Ciarlet [9].

Exercises. (1) Find the eigenvalues and normalized eigenvectors of the matrix

$$A = \begin{pmatrix} 2 & 1 \\ 1 & 2 \end{pmatrix}.$$

Then use the normalized eigenvectors to construct the matrix Q^{-1} such that $Q^{-1}AQ$ is a diagonal matrix.

(2) Show that the matrix

$$\begin{pmatrix} 0 & 1 & 0 & 0 \\ 0 & 0 & 1 & 0 \\ 0 & 0 & 0 & 1 \\ 0 & 0 & 0 & 0 \end{pmatrix}$$

is not diagonalizable.

(3) Let O be an orthogonal matrix. Show that there exists an orthogonal matrix Q such that

$$Q^{-1}OQ = \begin{pmatrix} 1 &&&&&&&&&& \\ & \ddots &&&&&&&&& \\ && 1 &&&&&&&& \\ &&& -1 &&&&&&& \\ &&&& \ddots &&&&&& \\ &&&&& -1 &&&&& \\ &&&&&& \cos\theta_1 & \sin\theta_1 &&& \\ &&&&&& -\sin\theta_1 & \cos\theta_1 &&& \\ &&&&&&&& \ddots && \\ &&&&&&&&& \cos\theta_r & \sin\theta_r \\ &&&&&&&&& -\sin\theta_r & \cos\theta_r \end{pmatrix}.$$

(4) Let A be a real matrix of order n. Show that a necessary and sufficient condition for the existence of a unitary matrix U of the same order and of a real matrix B (of the same order) such that $U = A + iB$ (in other word, such that the matrix A is the 'real part' of the matrix U) is that all the singular values of the matrix A should be not greater than 1.

1.7 Permutation Matrices

In this section we introduce permutation matrices and discuss their properties. The connection with the Kronecker product is described in section 2.4. By a *permutation* $\boldsymbol{\sigma}$ of the set

$$N := \{1,\ 2,\ \ldots,\ n\}$$

is meant a one-to-one mapping of N onto itself. Including the identity permutation there are $n!$ distinct permutations of N. We indicate a permutation by

$$\boldsymbol{\sigma}(1) = i_1$$
$$\boldsymbol{\sigma}(2) = i_2$$
$$\vdots$$
$$\boldsymbol{\sigma}(n) = i_n$$

which is written as

$$\boldsymbol{\sigma} : \begin{pmatrix} 1 & 2 & \cdots & n \\ i_1 & i_2 & \cdots & i_n \end{pmatrix}.$$

The inverse permutation is designated by $\boldsymbol{\sigma}^{-1}$. Thus

$$\boldsymbol{\sigma}^{-1}(i_k) = k.$$

Let $\mathbf{e}_j^T$ denote the unit (row) vector of n components which has a 1 in the j-th position and 0's elsewhere

$$\mathbf{e}_j^T := (0, \cdots, 0, 1, 0, \cdots, 0).$$

Definition. By a *permutation matrix* of order n is meant a matrix of the form

$$P = P_\sigma = \begin{pmatrix} \mathbf{e}_{i_1}^T \\ \mathbf{e}_{i_2}^T \\ \vdots \\ \mathbf{e}_{i_n}^T \end{pmatrix}.$$

The i-th row of P has a 1 in the $\boldsymbol{\sigma}(i)$-th column and 0's elsewhere. The j-th column of P has a 1 in the $\boldsymbol{\sigma}^{-1}(j)$-th row and 0's elsewhere. Thus each row and each column of P has precisely one 1 in it. It is easily seen that

$$P_\sigma \begin{pmatrix} x_1 \\ x_2 \\ \vdots \\ x_n \end{pmatrix} = \begin{pmatrix} x_{\sigma(1)} \\ x_{\sigma(2)} \\ \vdots \\ x_{\sigma(n)} \end{pmatrix}.$$

Example. Let

$$\boldsymbol{\sigma} : \begin{pmatrix} 1 & 2 & 3 & 4 \\ 4 & 1 & 3 & 2 \end{pmatrix}.$$

Then

$$P_\sigma = \begin{pmatrix} 0 & 0 & 0 & 1 \\ 1 & 0 & 0 & 0 \\ 0 & 0 & 1 & 0 \\ 0 & 1 & 0 & 0 \end{pmatrix}. \quad \clubsuit$$

Example. The set of all 3×3 permutation matrices are given by

$$\left\{ \begin{pmatrix} 1 & 0 & 0 \\ 0 & 1 & 0 \\ 0 & 0 & 1 \end{pmatrix} \quad \begin{pmatrix} 1 & 0 & 0 \\ 0 & 0 & 1 \\ 0 & 1 & 0 \end{pmatrix} \quad \begin{pmatrix} 0 & 1 & 0 \\ 1 & 0 & 0 \\ 0 & 0 & 1 \end{pmatrix} \right.$$

$$\left. \begin{pmatrix} 0 & 1 & 0 \\ 0 & 0 & 1 \\ 1 & 0 & 0 \end{pmatrix} \quad \begin{pmatrix} 0 & 0 & 1 \\ 1 & 0 & 0 \\ 0 & 1 & 0 \end{pmatrix} \quad \begin{pmatrix} 0 & 0 & 1 \\ 0 & 1 & 0 \\ 1 & 0 & 0 \end{pmatrix} \right\}. \quad \clubsuit$$

We can easily be proved that

$$P_\sigma P_\tau = P_{\sigma\tau}$$

where the product of the permutations $\boldsymbol{\sigma}$, $\boldsymbol{\tau}$ is applied from left to right. Furthermore,

$$(P_\sigma)^* = P_{\sigma^{-1}}.$$

Hence

$$(P_\sigma)^* P_\sigma = P_{\sigma^{-1}} P_\sigma = P_I = I$$

where I is the $n \times n$ unit matrix. It follows that

$$(P_\sigma)^* = P_{\sigma^{-1}} = (P_\sigma)^{-1}.$$

Consequently, the permutation matrices form a group under matrix multiplication. We find that the permutation matrices are unitary, forming a finite subgroup of the unitary group (see section 1.11 for more details on group theory).

Exercises. (1) Show that the number of $n \times n$ permutation matrices is given by $n!$.

(2) Find all 4×4 permutation matrices.

(3) Show that the determinant of a permutation matrix is either $+1$ or -1.

(4) Show that the eigenvalues λ_j of a permutation matrix are $\lambda_j \in \{1, -1\}$.

(5) Show that the rank of an $n \times n$ permutation matrix is n.

(6) Consider the set of all $n \times n$ permutation matrices. How many of the elements are their own inverses, i.e. $P = P^{-1}$?

(7) Consider the 4×4 permutation matrix

$$P = \begin{pmatrix} 0 & 0 & 0 & 1 \\ 0 & 0 & 1 & 0 \\ 0 & 1 & 0 & 0 \\ 1 & 0 & 0 & 0 \end{pmatrix}.$$

Find all the eigenvalues and normalized eigenvectors. From the normalized eigenvectors construct an invertible matrix Q such that $Q^{-1}PQ$ is a diagonal matrix.

(8) Let P_1 and P_2 be two $n \times n$ permutation matrices. Is $[P_1, P_2] = 0$, where $[\,,\,]$ denotes the commutator ? The commutator is defined by $[P_1, P_2] := P_1P_2 - P_2P_1$.

(9) Is it possible to find $\mathbf{v} \in \mathbf{R}^n$ such that $\mathbf{v}\mathbf{v}^T$ is an $n \times n$ permutation matrix ?

1.8 Vec Operator

Let A be an $m \times n$ matrix. A matrix operation is that of stacking the columns of a matrix one under the other to form a single column. This operation is called vec (Neudecker [25], Brewer [8], Graham [14], Searle [30]). Thus vec$A$ is a vector of order $m \times n$.

Example. Let

$$A = \begin{pmatrix} 1 & 2 & 3 \\ 4 & 5 & 6 \end{pmatrix}.$$

Then

$$\text{vec}A = \begin{pmatrix} 1 \\ 4 \\ 2 \\ 5 \\ 3 \\ 6 \end{pmatrix}. \qquad \clubsuit$$

The notations vec$A$ and vec(A) are used interchangeably, the latter only when parentheses are necessary. Let A, B be $m \times n$ matrices. We can easily prove that

$$\text{vec}(A + B) = \text{vec}A + \text{vec}B.$$

This means the vec-operation is linear. It is also easy to see that

$$\text{vec}(\alpha A) = \alpha \text{vec}(A), \qquad \alpha \in \mathbf{C}.$$

An extension of vec$A$ is vech$A$, defined in the same way that vec$A$ is, except that for each column of A only that part of it which is on or below the diagonal of A is put into vech$A$ (vector-half of A). In this way, for A symmetric, vech$A$ contains only the distinct elements of A.

Example.

$$A = \begin{pmatrix} 1 & 7 & 6 \\ 7 & 3 & 8 \\ 6 & 8 & 2 \end{pmatrix} = A^T$$

$$\text{vech}A = \begin{pmatrix} 1 \\ 7 \\ 6 \\ 3 \\ 8 \\ 2 \end{pmatrix}. \qquad \clubsuit$$

The following theorems give useful properties of the vec operator. Proofs of the first depend on the elementary vector $\mathbf{e}_j$, the j-th column of a unit matrix, i. e.

$$\mathbf{e}_1 = \begin{pmatrix} 1 \\ 0 \\ \vdots \\ 0 \end{pmatrix}, \qquad \mathbf{e}_2 = \begin{pmatrix} 0 \\ 1 \\ \vdots \\ 0 \end{pmatrix}, \ldots, \mathbf{e}_n = \begin{pmatrix} 0 \\ 0 \\ \vdots \\ 1 \end{pmatrix}$$

and

$$\mathbf{e}_1^T = (1, 0, \ldots, 0), \qquad \mathbf{e}_2^T = (0, 1, \ldots, 0), \ \ldots, \ \mathbf{e}_n^T = (0, \ldots, 0, 1).$$

Theorem. Let A, B be $n \times n$ matrices. Then

$$\mathrm{tr}(AB) = (\mathrm{vec}A^T)^T \mathrm{vec}B.$$

Proof.

$$\mathrm{tr}(AB) = \sum_{i=1}^{n} \mathbf{e}_i^T AB\mathbf{e}_i = (\mathbf{e}_1^T A \cdots \mathbf{e}_r^T A) \begin{pmatrix} B\mathbf{e}_1 \\ B\mathbf{e}_2 \\ \vdots \\ B\mathbf{e}_r \end{pmatrix} = (\mathrm{vec}A^T)^T \mathrm{vec}B. \quad \spadesuit$$

Theorem. Let A be an $m \times m$ matrix. Then there is a permutation matrix P such that

$$\mathrm{vec}(A) = P\mathrm{vec}(A^T).$$

The proof is left to the reader as an exercise.

Example. Let

$$A = \begin{pmatrix} 1 & 2 \\ 3 & 4 \end{pmatrix}, \qquad A^T = \begin{pmatrix} 1 & 3 \\ 2 & 4 \end{pmatrix}.$$

Then $\mathrm{vec}(A) = P\mathrm{vec}(A^T)$, where

$$P = \begin{pmatrix} 1 & 0 & 0 & 0 \\ 0 & 0 & 1 & 0 \\ 0 & 1 & 0 & 0 \\ 0 & 0 & 0 & 1 \end{pmatrix}.$$

The full power of the vec operator will be seen when we consider the Kronecker product and the vec operator (see section 2.11).

1.9 Vector and Matrix Norms

Definition. Let V be an n-dimensional vector space over the field K of scalars. A *norm* on V is a function $||\cdot|| : V \to \mathbf{R}$ which satisfies the following properties:

$$||\mathbf{v}|| = 0 \Leftrightarrow \mathbf{v} = 0, \text{ and } ||\mathbf{v}|| \geq 0 \text{ for every } \mathbf{v} \in V$$

$$||\alpha \mathbf{v}|| = |\alpha| ||\mathbf{v}|| \text{ for every } \alpha \in K \text{ and } \mathbf{v} \in V$$

$$||\mathbf{u} + \mathbf{v}|| \leq ||\mathbf{u}|| + ||\mathbf{v}|| \text{ for every } \mathbf{u}, \mathbf{v} \in V.$$

The last property is known as the *triangle inequality.* A norm on V will also be called a *vector norm.* . We call a vector space which is provided with a norm a *normed vector space.*

Let V be a finite-dimensional space. The following three norms are the ones most commonly used in practice

$$||\mathbf{v}||_1 := \sum_{i=1}^{n} |v_i|$$

$$||\mathbf{v}||_2 := \left(\sum_{i=1}^{n} |v_i|^2 \right)^{1/2} = (\mathbf{v}, \mathbf{v})^{1/2}$$

$$||\mathbf{v}||_\infty := \max_{1 \leq i \leq n} |v_i|$$

the norm $||\cdot||_2$ is called the *Euclidian norm.* It is easy to verify directly that the two functions $||\cdot||_1$ and $||\cdot||_\infty$ are indeed norms. As for the function $||\cdot||_2$, it is a particular case of the following more general result.

Theorem. Let V be a finite-dimensional vector space. For every real number $p \geq 1$, the function $||\cdot||_p$ defined by

$$||\mathbf{v}||_p := \left(\sum_{i=1}^{n} |v_i|^p \right)^{1/p}$$

is a norm.

For the proof refer to Ciarlet [9].

The proof uses the following inequalities: For $p > 1$ and $1/p + 1/q = 1$, the inequality

$$\sum_{i=1}^{n} |u_i v_i| \leq \left(\sum_{i=1}^{n} |u_i|^p \right)^{1/p} \left(\sum_{i=1}^{n} |v_i|^q \right)^{1/q}$$

is called *Hölder's inequality.* Hölder's inequality for $p = 2$,

$$\sum_{i=1}^{n} |u_i v_i| \leq \left(\sum_{i=1}^{n} |u_i|^2 \right)^{1/2} \left(\sum_{i=1}^{n} |v_i|^2 \right)^{1/2}$$

is called *the Cauchy-Schwarz inequality.* The triangle inequality for the norm $||\cdot||_p$,

$$\left(\sum_{i=1}^{n}|u_i+v_i|^p\right)^{1/p} \le \left(\sum_{i=1}^{n}|u_i|^p\right)^{1/p} + \left(\sum_{i=1}^{n}|v_i|^p\right)^{1/p}$$

is called *Minkowski's inequality.*

The norms defined above are *equivalent*, this property being a particular case of the equivalence of norms in a finite-dimensional space.

Definition. Two norms $||\cdot||$ and $||\cdot||'$, defined over the same vector space V, are equivalent if there exist two constants C and C' such that

$$||\mathbf{v}||' \le C||\mathbf{v}|| \quad \text{and} \quad ||\mathbf{v}|| \le C'||\mathbf{v}||' \quad \text{for every} \quad \mathbf{v} \in V.$$

Let $\mathcal{A}_n$ be the ring of matrices of order n, with elements in the field K.

Definition. A *matrix norm* is a function $||\cdot|| : \mathcal{A}_n \to \mathbf{R}$ which satisfies the following properties

$$||A|| = 0 \Leftrightarrow A = 0 \text{ and } ||A|| \ge 0 \text{ for every } A \in \mathcal{A}_n$$
$$||\alpha A|| = |\alpha|||A|| \text{ for every } \alpha \in K, \quad A \in \mathcal{A}_n$$
$$||A+B|| \le ||A|| + ||B|| \text{ for every } A, B \in \mathcal{A}_n$$
$$||AB|| \le ||A||||B|| \text{ for every } A, B \in \mathcal{A}_n.$$

The ring $\mathcal{A}_n$ is itself a vector space of dimension n^2. Thus the first three properties above are nothing other than those of a vector norm, considering a matrix as a vector with n^2 components. The last property is evidently special to square matrices.

The result which follows gives a particularly simple means of constructing matrix norms.

Definition. Given a vector norm $||\cdot||$ on $\mathbf{C}^n$, the function $||\cdot|| : \mathcal{A}_n(\mathbf{C}) \to \mathbf{R}$ defined by

$$||A|| := \sup_{\substack{\mathbf{v}\in\mathbf{C}^n \\ ||\mathbf{v}||=1}} ||A\mathbf{v}||$$

is a matrix norm, called the *subordinate matrix norm* (subordinate to the given vector norm). Sometimes it is also called the *induced matrix norm.*

This is just one particular case of the usual definition of the norm of a linear transformation.

Example. Consider the matrix

$$A = \begin{pmatrix} 1 & 1 \\ 2 & 2 \end{pmatrix}.$$

Then we find

$$||A|| = \sup_{\substack{\mathbf{v}\in\mathbf{C}^n \\ ||\mathbf{v}||=1}} ||A\mathbf{v}|| = \sqrt{10}.$$

This result can be found by using the method of the *Lagrange multiplier*. The constraint is $||\mathbf{v}|| = 1$. Furthermore we note that the eigenvalues of the matrix

$$A^T A = \begin{pmatrix} 1 & 2 \\ 1 & 2 \end{pmatrix} \begin{pmatrix} 1 & 1 \\ 2 & 2 \end{pmatrix} = \begin{pmatrix} 5 & 5 \\ 5 & 5 \end{pmatrix}$$

are given by $\lambda_1 = 10$ and $\lambda_2 = 0$. Thus the norm of A is the square root of the largest eigenvalue of $A^T A$.

It follows from the definition of a subordinate norm that

$$||A\mathbf{v}|| \leq ||A||||\mathbf{v}|| \quad \text{for every } \mathbf{v} \in \mathbf{C}^n.$$

We note that a subordinate norm always satisfies

$$||I|| = 1$$

where I is the the $n \times n$ unit matrix. Let us now calculate each of the subordinate norms of the vector norms $||\cdot||_1, ||\cdot||_2, ||\cdot||_\infty$.

Theorem. Let $A = (a_{ij})$ be a square matrix. Then

$$||A||_1 := \sup_{||\mathbf{v}||=1} ||A\mathbf{v}||_1 = \max_{1\leq j\leq n} \sum_{i=1}^{n} |a_{ij}|$$

$$||A||_2 := \sup_{||\mathbf{v}||=1} ||A\mathbf{v}||_2 = \sqrt{\varrho(A^*A)} = ||A^*||_2$$

$$||A||_\infty := \sup_{||\mathbf{v}||=1} ||A\mathbf{v}||_\infty = \max_{1\leq i\leq n} \sum_{j=1}^{n} |a_{ij}|$$

where $\varrho(A^*A)$ is the spectral radius of A^*A. The norm $||\cdot||_2$ is invariant under unitary transformations

$$UU^* = I \Rightarrow ||A||_2 = ||AU||_2 = ||UA||_2 = ||U^*AU||_2.$$

Furthermore, if the matrix A is normal we have

$$AA^* = AA^* \Rightarrow ||A||_2 = \varrho(A).$$

The invariance of the norm $||\cdot||_2$ under unitary transformations is nothing more than the interpretation of the equalities

$$\varrho(A^*A) = \varrho(U^*A^*AU) = \varrho(A^*U^*UA) = \varrho(U^*A^*UU^*AU).$$

If the matrix A is normal, there exists a unitary matrix U such that

$$U^*AU = \mathrm{diag}(\lambda_i(A)) := D.$$

Accordingly,

$$A^*A = (UDU^*)^*UDU^* = UD^*DU^*$$

which proves that

$$\varrho(A^*A) = \varrho(D^*D) = \max_{1\leq i\leq n} |\lambda_i(A)|^2 = (\varrho(A))^2.$$

The norm $||A||_2$ is nothing other than the largest singular value of the matrix A. If a matrix A is Hermitian, or symmetric (and hence normal), we have $||A||_2 = \varrho(A)$. If a matrix A is unitary, or orthogonal (and hence normal), we have

$$||A||_2 = \sqrt{\varrho(A^*A)} = \sqrt{\varrho(I)} = 1.$$

There exist matrix norms which are not subordinate to any vector norm. An example of a matrix norm which is not subordinate is given in the following theorem.

Theorem. The function $||\cdot||_E : \mathcal{A}_n \to \mathbf{R}$ defined by

$$||A||_E = \left(\sum_{i=1}^{n}\sum_{j=1}^{n} |a_{ij}|^2\right)^{1/2} = (\mathrm{tr}(A^*A))^{1/2}$$

for every matrix $A = (a_{ij})$ of order n is a matrix norm which is not subordinate (for $n \geq 2$). Furthermore, the function is invariant under unitary transformations,

$$UU^* = I \Rightarrow ||A||_E = ||AU||_E = ||UA||_E = ||U^*AU||_E$$

and satisfies

$$||A||_2 \leq ||A||_E \leq \sqrt{n}||A||_2 \text{ for every } A \in \mathcal{A}_n.$$

For the proof refer to Ciarlet [9].

Example. Let I be the unit matrix. Then

$$||I||_E = \sqrt{2}.$$

Exercises. (1) Let A, B be $n \times n$ matrices over $\mathbf{C}$. Show that

$$(A, B) := \operatorname{tr}(AB^*)$$

defines a scalar product. Then

$$||A|| = \sqrt{(A, A)}$$

defines a norm. Find $||A||$ for

$$A = \begin{pmatrix} 0 & i \\ -i & 0 \end{pmatrix}.$$

(2) Given a diagonalizable matrix A, does a matrix norm $||\cdot||$ exist for which $\varrho(A) = ||A||$?

(3) Let $\mathbf{v} = \exp(i\alpha)\mathbf{u}$, where $\alpha \in \mathbf{R}$. Show that $||\mathbf{u}|| = ||\mathbf{v}||$.

(4) What can be said about the norm of a nilpotent matrix ?

(5) Let A be a Hermitian matrix. Find a necessary and sufficient condition for the function $\mathbf{v} \to (\mathbf{v}^* A\mathbf{v})^{1/2}$ to be a norm.

(6) Let $||\cdot||$ be a subordinate matrix norm and A an $n \times n$ matrix satisfying $||A|| < 1$. Show that the matrix $I_n + A$ is invertible and

$$||(I_n + A)^{-1}|| \le \frac{1}{1 - ||A||}.$$

(7) Prove that the function

$$\mathbf{v} \in \mathbf{C}^n \to ||\mathbf{v}||_p = \left(\sum_{i=1}^{n} |v_i|^p \right)^{1/p}$$

is not a norm when $0 < p < 1$ (unless $n = 1$).

(8) Find the smallest constants C for which

$$||\mathbf{v}|| \le C||\mathbf{v}||' \quad \text{for every } \mathbf{v} \in K^n$$

when the distinct norms $||\cdot||$ and $||\cdot||'$ are chosen from the set $\{||\cdot||_1, ||\cdot||_2, ||\cdot||_\infty\}$.

1.10 Sequences of Vectors and Matrices

Definition. In a vector space V, equipped with a norm $||\cdot||$, a sequence (x_k) of elements of V is said to converge to an element $x \in V$, which is the limit of the sequence (x_k), if

$$\lim_{k\to\infty} ||x_k - x|| = 0$$

and one writes

$$x = \lim_{k\to\infty} x_k.$$

If the space is finite-dimensional, the equivalence of the norms shows that the convergence of a sequence is independent of the norm chosen. The particular choice of the norm $||\cdot||_\infty$ shows that the convergence of a sequence of vectors is equivalent to the convergence of n sequences (n being equal to the dimension of the space) of scalars consisting of the components of the vectors.

Example. Let $V = \mathbf{C}^2$ and

$$\mathbf{u}_k := \begin{pmatrix} \exp(-k) \\ 1/(1+k) \end{pmatrix}, \qquad k = 1, 2, \ldots .$$

Then

$$\mathbf{u} = \lim_{k\to\infty} \mathbf{u}_k = \begin{pmatrix} 0 \\ 0 \end{pmatrix}.$$

By considering the set $\mathcal{A}_{m,n}(K)$ of matrices of type (m,n) as a vector space of dimension mn, one sees in the same way that the convergence of a sequence of matrices of type (m,n) is independent of the norm chosen, and that it is equivalent to the convergence of mn sequences of scalars consisting of the elements of these matrices. The following result gives necessary and sufficient conditions for the convergence of the particular sequence consisting of the successive powers of a given (square) matrix to the null matrix. From these conditions can be derived the fundamental criterion for the convergence of iterative methods for the solution of linear systems of equations.

Theorem. Let B be a square matrix. The following conditions are equivalent:

(1) $$\lim_{k\to\infty} B^k = 0,$$

(2) $$\lim_{k\to\infty} B^k \mathbf{v} = 0 \text{ for every vector } \mathbf{v}$$

(3) $$\varrho(B) < 1$$

(4) $$||B|| < 1 \text{ for at least one subordinate matrix norm } ||\cdot||.$$

For the proof of the theorem refer to Ciarlet [9].

Example. Consider the matrix

$$B = \begin{pmatrix} \frac{1}{4} & \frac{1}{4} \\ \frac{1}{4} & \frac{1}{4} \end{pmatrix}.$$

Then

$$\lim_{k \to \infty} B^k = 0 . \quad \clubsuit$$

The following theorem (Ciarlet [9]) is useful for the study of iterative methods, as regards the rate of convergence.

Theorem. Let A be a square matrix and let $|| \cdot ||$ be any subordinate matrix norm. Then

$$\lim_{k \to \infty} ||A^k||^{1/k} = \varrho(A).$$

Example. Let $A = I$, where I is the $n \times n$ unit matrix. Then $I^k = I$ and therefore $||I^k||^{1/k} = 1$. Moreover $\varrho(A) = 1$. $\clubsuit$

In theoretical physics and in particular in quantum mechanics a very important role is played by the exponential function of a square matrix. Let A be a square matrix. We set

$$A_k := I + \frac{A}{1!} + \frac{A^2}{2!} + \ldots + \frac{A^k}{k!}, \qquad k \geq 1.$$

The sequence (A_k) converges. Its limit is denoted by $\exp(A)$. We have

$$\exp(A) := \sum_{k=0}^{\infty} \frac{A^k}{k!}.$$

Let ϵ be a parameter ($\epsilon \in \mathbf{R}$). Then we obtain

$$\exp(\epsilon A) = \sum_{k=0}^{\infty} \frac{(\epsilon A)^k}{k!}.$$

Theorem. Let A be an $n \times n$ matrix. Then

$$\det(\exp(A)) \equiv \exp(\operatorname{tr}(A)).$$

For the proof refer to Steeb [37]. The theorem shows that the matrix $\exp(A)$ is always invertible. If A is the zero matrix, then we have $\exp(A) = I$, where I is the identity matrix.

Exercises. (1) Let

$$A = \begin{pmatrix} 0 & 1 \\ 1 & 0 \end{pmatrix}.$$

Calculate $\exp(\epsilon A)$, where $\epsilon \in \mathbf{R}$.

(2) Let

$$A = \begin{pmatrix} 1 & 1 \\ 1 & 1 \end{pmatrix}.$$

Find $\cos(A)$.

(3) Let A be an $n \times n$ matrix such that $A^2 = I$, where I is the $n \times n$ unit matrix. Let $\epsilon \in \mathbf{R}$. Show that

$$\exp(\epsilon A) = I \cosh(\epsilon) + A \sinh(\epsilon).$$

(4) Let A be a square matrix such that the sequence $(A^k)_{k \geq 1}$ converges to an invertible matrix. Find A.

(5) Let B be a square matrix satisfying $||B|| < 1$. Prove that the sequence $(C_k)_{k \geq 1}$, where

$$C_k = I + B + B^2 + \cdots + B^k$$

converges and that

$$\lim_{k \to \infty} C_k = (I - B)^{-1}.$$

(6) Prove that

$$AB = BA \Rightarrow \exp(A + B) = \exp(A) \exp(B).$$

(7) Let (A_k) be a sequence of $n \times n$ matrices. Show that the following conditions are equivalent:
(i) the sequence (A_k) converges;
(ii) for every vector $\mathbf{v} \in \mathbf{R}^n$, the sequence of vectors $(A_k \mathbf{v})$ converges in $\mathbf{R}^n$.

(8) Let A and B be square matrices. Assume that $\exp(A)\exp(B) = \exp(A + B)$. Show that in general $[A, B] \neq 0$.

(9) Extend the Taylor expansion for $\ln(1 + x)$

$$\ln(1 + x) = x - \frac{x^2}{2} + \frac{x^3}{3} - \frac{x^4}{4} + \ldots \qquad -1 < x \leq 1$$

to $n \times n$ matrices.

1.11 Groups

In the representation of groups as $n \times n$ matrices the Kronecker product plays a central role. We give a short introduction to group theory and then discuss the connection with the Kronecker product. For further reading in group theory we refer to the books of Miller [24], Baumslag and Chandler [6] and Steeb [36]. In sections 2.12 and 2.13 we give a more detailed introduction to representation theory and the connection with the Kronecker product.

Definition. A *group* G is a set of objects $\{g, h, k, \ldots\}$ (not necessarily countable) together with a binary operaton which associates with any ordered pair of elements $g, h \in G$ a third element $gh \in G$. The binary operation (called *group multiplication)* is subject to the following requirements:
(1) There exists an element e in G called the *identity element* such that $ge = eg = g$ for all $g \in G$.
(2) For every $g \in G$ there exists in G an *inverse element* g^{-1} such that $gg^{-1} = g^{-1}g = e$.
(3) *Associative law.* The identity $(gh)k = g(hk)$ is satisfied for all $g, h, k \in G$.

Thus, any set together with a binary operation which satisfies conditions (1) - (3) is called a group.

If $gh = hg$ we say that the elements g and h *commute.* If all elements of G commute then G is a *commutative* or *abelian* group. If G has a finite number of elements it has *finite order* $n(G)$, where $n(G)$ is the number of elements. Otherwise, G has *infinite order.*

A *subgroup* H of G is a subset which is itself a group under the group multiplication defined in G. The subgroups G and $\{e\}$ are called *improper* subgroups of G. All other subgroups are *proper.*

If a group G consists of a finite number of elements, then G is called a *finite group*; otherwise, G is called an *infinite group.*

Example. The set of integers $\mathbf{Z}$ with addition as group composition is an infinite additive group with $e = 0$. ♣

Example. The set $\{1, -1\}$ with multiplication as group composition is a finite abelian group with $e = 1$. ♣

Definition. Let G be a finite group. The number of elements of G is called the *dimension* or *order* of G.

Theorem. The order of a subgroup of a finite group divides the order of the group.

This theorem is called *Lagrange's theorem.* For the proof we refer to the literature (Miller [24]).

A way to partition G is by means of *conjugacy classes.*

Definition. A group element h is said to be conjugate to the group element $k, h \sim k$, if there exists a $g \in G$ such that

$$k = ghg^{-1}.$$

It is easy to show that conjugacy is an equivalence relation, i.e., (1) $h \sim h$ (reflexive), (2) $h \sim k$ implies $k \sim h$ (symmetric), and (3) $h \sim k, k \sim j$ implies $h \sim j$ (transitive). Thus, the elements of G can be divided into *conjugacy classes* of mutually conjugate elements. The class containing e consists of just one element since

$$geg^{-1} = e$$

for all $g \in G$. Different conjugacy classes do not necessarily contain the same number of elements.

Let G be an abelian group. Then each conjugacy class consists of one group element each, since

$$ghg^{-1} = h, \quad \text{for all} \quad g \in G.$$

Let us now give a number of examples to illustrate the definitions given above.

Example. A *field* is an (infinite) abelian group with respect to addition. The set of nonzero elements of a field forms a group with respect to multiplication, which is called a multiplicative group of the field. ♣

Example. A *linear vector space* over a field K (such as the real numbers $\mathbf{R}$) is an abelian group with respect to the usual addition of vectors. The group composition of two elements (vectors) $\mathbf{a}$ and $\mathbf{b}$ is their vector sum $\mathbf{a} + \mathbf{b}$. The identity element is the zero vector and the inverse of an element is its negative. ♣

Example. Let N be an integer with $N \geq 1$. The set

$$\{ e^{2\pi i n/N} \quad : n = 0, 1, \ldots, N-1 \}$$

is an abelian (finite) group under multiplication since

$$\exp(2\pi i n/N) \exp(2\pi i m/N) = \exp(2\pi i (n+m)/N)$$

where $n, m = 0, 1, \ldots, N-1$. Note that $\exp(2\pi i n) = 1$ for $n \in \mathbf{N}$. We consider some special cases of N: For $N = 2$ we find the set $\{1, -1\}$ and for $N = 4$ we find

$\{1, i, -1, -i\}$. These are elements on the unit circle in the complex plane. For $N \to \infty$ the number of points on the unit circle increases. As $N \to \infty$ we find the unitary group

$$U(1) := \left\{ e^{i\alpha} \; : \; \alpha \in \mathbf{R} \right\}. \quad ♣$$

Example. The two matrices

$$\left\{ \begin{pmatrix} 1 & 0 \\ 0 & 1 \end{pmatrix}, \begin{pmatrix} 0 & 1 \\ 1 & 0 \end{pmatrix} \right\}$$

form a finite abelian group of order two with matrix multiplication as group composition. The closure can easily be verified

$$\begin{pmatrix} 1 & 0 \\ 0 & 1 \end{pmatrix}\begin{pmatrix} 1 & 0 \\ 0 & 1 \end{pmatrix} = \begin{pmatrix} 1 & 0 \\ 0 & 1 \end{pmatrix}, \quad \begin{pmatrix} 1 & 0 \\ 0 & 1 \end{pmatrix}\begin{pmatrix} 0 & 1 \\ 1 & 0 \end{pmatrix} = \begin{pmatrix} 0 & 1 \\ 1 & 0 \end{pmatrix}$$

$$\begin{pmatrix} 0 & 1 \\ 1 & 0 \end{pmatrix}\begin{pmatrix} 0 & 1 \\ 1 & 0 \end{pmatrix} = \begin{pmatrix} 1 & 0 \\ 0 & 1 \end{pmatrix}.$$

The identity element is the 2×2 unit matrix. ♣

Example. Let $M = \{1, 2, \ldots, n\}$. Let $Bi(M, M)$ be the set of bijective mappings $\sigma : M \to M$ so that

$$\sigma : \{1, 2, \ldots, n\} \to \{p_1, p_2, \ldots, p_n\}$$

forms a group S_n under the composition of functions. Let S_n be the set of all the permutations

$$\sigma = \begin{pmatrix} 1 & 2 & \cdots & n \\ p_1 & p_2 & \cdots & p_n \end{pmatrix}.$$

We say 1 is mapped into p_1, 2 into p_2, $\ldots$, n into p_n. The numbers $p_1, p_2, \ldots, p_n$ are a reordering of $1, 2, \ldots, n$ and no two of the p_j's $j = 1, 2 \ldots, n$ are the same. The inverse permutation is given by

$$\sigma^{-1} = \begin{pmatrix} p_1 & p_2 & \cdots & p_n \\ 1 & 2 & \cdots & n \end{pmatrix}.$$

The product of two permutations σ and τ, with

$$\tau = \begin{pmatrix} q_1 & q_2 & \cdots & q_n \\ 1 & 2 & \cdots & n \end{pmatrix}$$

is given by the permutation

$$\sigma \circ \tau = \begin{pmatrix} q_1 & q_2 & \cdots & q_n \\ p_1 & p_2 & \cdots & p_n \end{pmatrix}.$$

That is, the integer q_i is mapped to i by τ and i is mapped to p_i by σ, so q_i is mapped to p_i by $\sigma \circ \tau$. The identity permutation is

$$e = \begin{pmatrix} 1 & 2 & \cdots & n \\ 1 & 2 & \cdots & n \end{pmatrix}.$$

S_n has order $n!$. The group of all permutations on M is called the *symmetric group* on M which is non-abelian, if $n > 2$. ♣

Example. Let N be a positive integer. The set of all matrices

$$Z_{2\pi k/N} = \begin{pmatrix} \cos\frac{2k\pi}{N} & -\sin\frac{2k\pi}{N} \\ \sin\frac{2k\pi}{N} & \cos\frac{2k\pi}{N} \end{pmatrix}$$

where $k = 0, 1, 2, \ldots, N-1$, forms an abelian group under matrix multiplication. The elements of the group can be generated from the transformation

$$Z_{2k\pi/N} = \left(Z_{2\pi/N}\right)^k, \qquad k = 0, 1, 2, \ldots, N-1.$$

For example, if $N = 2$ the group consists of the elements $\{(Z_\pi)^0, (Z_\pi)^1\} \equiv \{-I, +I\}$ where I is the 2×2 unit matrix. This is an example of a cyclic group. ♣

Example. The set of all invertible $n \times n$ matrices form a group with respect to the usual multiplication of matrices. The group is called the *general linear group* over the real numbers $GL(n, \mathbf{R})$, or over the complex numbers $GL(n, \mathbf{C})$. This group together with its subgroups are the so-called *classical groups* which are Lie groups. ♣

Example. Let $\mathbf{C}$ be the complex plane. Let $z \in \mathbf{C}$. The set of Möbius transformations in $\mathbf{C}$ form a group called the *Möbius group* denoted by M where $m : \mathbf{C} \to \mathbf{C}$,

$$M := \{\, m(a,b,c,d) \ : \ a,b,c,d \in \mathbf{C}, \ \ ad - bc \neq 0 \,\}$$

and

$$m : z \mapsto z' = \frac{az+b}{cz+d}.$$

The condition $ad - bc \neq 0$ must hold for the transformation to be invertible. Here, $z = x + iy$, where $x, y \in \mathbf{R}$. This forms a group under the composition of functions: Let

$$m(z) = \frac{az+b}{cz+d}, \qquad \widetilde{m}(z) = \frac{ez+f}{gz+h}$$

where $ad - bc \neq 0$ and $eh - fg \neq 0$ ($e, f, g, h \in \mathbf{C}$). Consider the composition

$$\begin{aligned} m\left(\widetilde{m}(z)\right) &= \frac{a(ez+f)/(gz+h)+b}{c(ez+f)/(gz+h)+d} \\ &= \frac{aez+af+bgz+hb}{cez+cf+dgz+hd} \\ &= \frac{(ae+bg)z+(af+hb)}{(ce+dg)z+(cf+hd)}. \end{aligned}$$

Thus $m(\widetilde{m}(z))$ has the form of a Möbius transformation, since

$$(ae+bg)(cf+hd)-(af+hb)(ce+dg)$$

$$=aecf+aehd+bgcf+bghd-afce-afdg-hbce-hbdg$$

$$=ad(eh-fg)+bc(gf-eh)$$

$$=(ad-bc)(eh-fg)\neq 0.$$

Thus we conclude that m is closed under composition. Associativity holds since we consider the multiplication of complex numbers. The identity element is given by

$$m(1,0,0,1)=z.$$

To find the inverse of $m(z)$ we assume that

$$m\left(\widetilde{m}(z)\right)=\frac{(ae+bg)z+(af+hb)}{(ce+dg)z+(cf+hd)}=z$$

so that

$$ae+bg=1,\qquad af+hb=0,\qquad ce+dg=0,\qquad cf+hd=1$$

and we find

$$e=\frac{d}{ad-bc},\qquad f=-\frac{b}{ad-bc},\qquad g=-\frac{c}{ad-bc},\qquad h=\frac{a}{ad-bc}.$$

The inverse is thus given by

$$(z')^{-1}=\frac{dz-b}{-cz+a}.\qquad \clubsuit$$

Example. Let $\mathbf{Z}$ be the abelian group of integers. Let E be the set of even integers. Obviously, E is an abelian group under addition and is a subgroup of $\mathbf{Z}$. Let C_2 be the cyclic group of order 2. Then

$$\mathbf{Z}/E\cong C_2\,.$$

♣

We denote the mapping between two groups by ρ and present the following definitions:

Definition. A mapping of a group G into another group G' is called a *homomorphism* if it preserves all combinatorial operations associated with the group G so that

$$\rho(a\cdot b)=\rho(a)*\rho(b)$$

$a,b\in G$ and $\rho(a)$, $\rho(b)\in G'$. Here $\cdot$ and $*$ are the group compositions in G and G', respectively.

Example. There is a homomorphism ρ from $GL(2, \mathbf{C})$ into the Möbius group M given by

$$\rho : \begin{pmatrix} a & b \\ c & d \end{pmatrix} \to m(z) = \frac{az+b}{cz+d}.$$

We now check that ρ is indeed a homomorphism: Consider

$$A = \begin{pmatrix} a & b \\ c & d \end{pmatrix}$$

where $a, b, c, d \in \mathbf{C}$ and $ad - bc \neq 0$. The matrices A form a group with matrix multiplication as group composition. We find

$$AB = \begin{pmatrix} a & b \\ c & d \end{pmatrix} \begin{pmatrix} e & f \\ g & h \end{pmatrix} = \begin{pmatrix} ae+bg & af+bh \\ ce+dg & cf+dh \end{pmatrix}$$

where $e, f, g, h \in \mathbf{C}$. Thus

$$\rho(AB) = \frac{(ae+bg)z+(af+bh)}{(ce+dg)z+(cf+dh)}$$

and

$$\rho(A) = \frac{az+b}{cz+d}, \qquad \rho(B) = \frac{ez+f}{gz+h}$$

so that

$$\rho(A) * \rho(B) = \frac{(ae+bg)z+(af+bh)}{(ce+dg)z+(cf+dh)}.$$

We have shown that $\rho(A \cdot B) = \rho(A) * \rho(B)$ and thus that ρ is a homomorphism. ♣

An extension of the Möbius group is as follows. Consider the transformation

$$\mathbf{v} = \frac{A\mathbf{w}+B}{C\mathbf{w}+D}$$

where $\mathbf{v} = (v_1, \ldots, v_n)^T$, $\mathbf{w} = (w_1, \ldots, w_n)^T$ (T transpose). A is an $n \times n$ matrix, B an $n \times 1$ matrix, C a $1 \times n$ matrix and D a 1×1 matrix. The $(n+1) \times (n+1)$ matrix

$$\begin{pmatrix} A & B \\ C & D \end{pmatrix}$$

is invertible.

Example. An $n \times n$ permutation matrix is a matrix that has in each row and each column precisely one 1. There are $n!$ permutation matrices. The $n \times n$ permutation matrices form a group under matrix multiplication. Consider the symmetric group S_n given above. It is easy to see that the two groups are isomorphic. *Cayley's theorem* tells us that every finite group is isomorphic to a subgroup (or the group itself) of these permutation matrices. The six 3×3 permutation matrices are given by

$$A = \begin{pmatrix} 1 & 0 & 0 \\ 0 & 1 & 0 \\ 0 & 0 & 1 \end{pmatrix}, \qquad B = \begin{pmatrix} 1 & 0 & 0 \\ 0 & 0 & 1 \\ 0 & 1 & 0 \end{pmatrix}, \qquad C = \begin{pmatrix} 0 & 1 & 0 \\ 1 & 0 & 0 \\ 0 & 0 & 1 \end{pmatrix}$$

$$D = \begin{pmatrix} 0 & 1 & 0 \\ 0 & 0 & 1 \\ 1 & 0 & 0 \end{pmatrix}, \qquad E = \begin{pmatrix} 0 & 0 & 1 \\ 1 & 0 & 0 \\ 0 & 1 & 0 \end{pmatrix}, \qquad F = \begin{pmatrix} 0 & 0 & 1 \\ 0 & 1 & 0 \\ 1 & 0 & 0 \end{pmatrix}.$$

We have

$$AA = A \quad AB = B \quad AC = C \quad AD = D \quad AE = E \quad AF = F$$

$$BA = B \quad BB = A \quad BC = D \quad BD = C \quad BE = F \quad BF = E$$

$$CA = C \quad CB = E \quad CC = A \quad CD = F \quad CE = B \quad CF = D$$

$$DA = D \quad DB = F \quad DC = B \quad DD = E \quad DE = A \quad DF = C$$

$$EA = E \quad EB = C \quad EC = F \quad ED = A \quad EE = D \quad EF = B$$

$$FA = F \quad FB = D \quad FC = E \quad FD = B \quad FE = C \quad FF = A.$$

For the inverse we find

$$A^{-1} = A, \quad B^{-1} = B, \quad C^{-1} = C, \quad D^{-1} = E, \quad E^{-1} = D, \quad F^{-1} = F.$$

The order of a finite group is the number of elements of the group. Thus our group has order 6. Lagrange's theorem tells us that the order of a subgroup of a finite group divides the order of the group. Thus the subgroups must have order 3, 2, 1. From the group table we find the subgroups

$$\{A, D, E\}$$

$$\{A, B\}, \qquad \{A, C\}, \qquad \{A, F\}$$

$$\{A\}$$

Cayley's theorem tells us that every finite group is isomorphic to a subgroup (or the group itself) of these permutation matrices. The *order of an element* $g \in G$ is the order of the cyclic subgroup generated by $\{g\}$, i.e. the smallest positive integer m such that

$$g^m = e$$

where e is the identity element of the group. The integer m divides the order of G. Consider, for example, the element D of our group. Then

$$D^2 = E, \qquad D^3 = A, \qquad A \ \text{ identity element.}$$

Thus $m = 3$. ♣

Excerises. (1) Show that all $n \times n$ permutation matrices form a group under matrix multiplication.

(2) Find all subgroups of the group of the 4×4 matrices. Apply Cayley's theorem.

(3) Consider the matrices

$$A(\alpha) := \begin{pmatrix} \cos\alpha & \sin\alpha \\ -\sin\alpha & \cos\alpha \end{pmatrix}, \qquad \alpha \in \mathbf{R}.$$

Show that these matrices form a group under matrix multiplication.

(4) Consider the matrices

$$B(\alpha) := \begin{pmatrix} \cosh\alpha & \sinh\alpha \\ \sinh\alpha & \cosh\alpha \end{pmatrix}, \qquad \alpha \in \mathbf{R}.$$

Show that these matrices form a group under matrix multiplication.

(5) Consider the matrices given in (3). Find

$$X = \frac{d}{d\alpha} A(\alpha) \Big|_{\alpha=0}.$$

Show that

$$\exp(\alpha X) = A(\alpha).$$

(6) Let S be the set of even integers. Show that S is a group under addition of integers.

(7) Let S be the set of real numbers of the form $a + b\sqrt{2}$, where $a, b \in \mathbf{Q}$ and are not simultaneously zero. Show that S is a group under the usual multiplication of real numbers.

1.12 Lie Algebras

Lie algebras (Humphreys [17]) play a central role in theoretical physics. They are also closly linked to Lie groups and Lie transformation groups (Steeb [36]). In this section we give the definition of a Lie algebra and some applciations.

Definition. A vector space L over a field F, with an operation $L \times L \to L$ denoted by

$$(x,y) \to [x,y]$$

and called the commutator of x and y, is called a *Lie algebra* over F if the following axioms are satisfied:

(L1)	The bracket operation is bilinear.
(L2)	$[x,x] = 0 \qquad \text{for all} \quad x \in L$
(L3)	$[x,[y,z]] + [y,[z,x]] + [z,[x,y]] = 0 \qquad (x,y,z \in L).$

Remark: Axiom (L3) is called the *Jacobi identity.*

Notice that (L1) and (L2), applied to $[x+y, x+y]$, imply anticommutativity: (L2')

$$[x,y] = -[y,x].$$

Conversely, if char$F \neq 2$ (for **R** and **C** we have char$F = 0$), then (L2') will imply (L2).

Definition. Let X and Y be $n \times n$-matrices. Then the *commutator* $[X,Y]$ of X and Y is defined as

$$[X,Y] := XY - YX.$$

The $n \times n$ matrices over **R** or **C** form a Lie algebra under the commutator. This means we have the following properties (X, Y, V, W $n \times n$ matrices and $c \in$ **C**)

$$\begin{aligned} [cX,Y] &= c[X,Y] \\ [X,cY] &= c[X,Y] \end{aligned}$$

$$[X,Y] = -[Y,X]$$

$$[X+Y, V+W] = [X,V] + [X,W] + [Y,V] + [Y,W]$$

and

$$[X,[Y,V]] + [V,[X,Y]] + [Y,[V,X]] = 0.$$

The last equation is called the *Jacobi identity*. .

Definition. Two Lie algebras L, L' are called *isomorphic* if there exists a vector space isomorphism $\phi : L \to L'$ satisfying

$$\phi([x,y]) = [\phi(x), \phi(y)]$$

for all $x, y \in L$.

The Lie algebra of all $n \times n$ matrices over $\mathbf{C}$ is also called $gl(n, \mathbf{C})$. A basis is given by the matrices

$$(E_{ij}), \qquad i, j = 1, 2, \ldots, n$$

where (E_{ij}) is the matrix having 1 in the (i,j) position and 0 elsewhere. Since

$$(E_{ij})(E_{kl}) = \delta_{jk}(E_{il})$$

it follows that the commutator is given by

$$[(E_{ij}), (E_{kl})] = \delta_{jk}(E_{il}) - \delta_{li}(E_{kj}) .$$

Thus the coefficients are all ± 1 or 0.

The classical Lie algebras are sub-Lie algebras of $gl(n, \mathbf{C})$. For example, $sl(n, \mathbf{R})$ is the Lie algebra with the condition

$$\mathrm{tr}(X) = 0$$

for all $X \in gl(n, \mathbf{R})$. Furthermore, $so(n, \mathbf{R})$ is the Lie algebra with the condition that

$$X^T = -X \quad \text{and} \quad \mathrm{tr}(X) = 0$$

for all $X \in so(n, \mathbf{R})$. For $n = 2$ a basis element is given by

$$X = \begin{pmatrix} 0 & 1 \\ -1 & 0 \end{pmatrix} .$$

For $n = 3$ we have a basis

$$X_1 = \begin{pmatrix} 0 & 0 & 0 \\ 0 & 0 & -1 \\ 0 & 1 & 0 \end{pmatrix}, \qquad X_2 = \begin{pmatrix} 0 & 0 & 1 \\ 0 & 0 & 0 \\ -1 & 0 & 0 \end{pmatrix}, \qquad X_3 = \begin{pmatrix} 0 & -1 & 0 \\ 1 & 0 & 0 \\ 0 & 0 & 0 \end{pmatrix} .$$

Exercises. (1) Show that the 2×2 matrices with trace zero form a Lie algebra under the commutator. Show that

$$\begin{pmatrix} 0 & 1 \\ 0 & 0 \end{pmatrix}, \quad \begin{pmatrix} 0 & 0 \\ 1 & 0 \end{pmatrix}, \quad \begin{pmatrix} 1 & 0 \\ 0 & -1 \end{pmatrix}$$

form a basis. Hint: Show that

$$\mathrm{tr}([A, B]) = 0$$

for any two $n \times n$ matrices.

(2) Find all Lie algebras with dimension 2.

(3) Show that the set of all diagonal matrices form a Lie algebra under the commutator.

(4) Do all Hermitian matrices form a Lie algebra under the commutator ?

(5) Do all skew-Hermitian matrices form a Lie algebra under the commutator ?

(6) An *automorphism* of L is an isomorphism of L onto itself. Let $L = sl(n, \mathbf{R})$. Show that if $g \in GL(n, \mathbf{R})$ and if

$$gLg^{-1} = L$$

then the map

$$x \mapsto gxg^{-1}$$

is an automorphism.

(7) The *center* of a Lie algebra L is defined as

$$Z(L) := \{\, z \in L \,:\, [z, x] = 0 \quad \text{for all}\, x \in L\,\}\,.$$

Find the center for the Lie algebra $sl(2, \mathbf{R})$.

1.13 Application in Quantum Theory

The dynamics in quantum mechanics is governed by the Schrödinger equation which describes the time evolution of the wave function (wave vector) and the Heisenberg equation of motion which describes the time evolution of operators. The quantum mechanical system is described by a Hamilton operator $\hat{H}$. We consider the special case that the Hamilton operator $\hat{H}$ acts in the finite dimensional vector space $\mathbf{C}^n$.

The *Schrödinger equation* is given by

$$i\hbar\frac{\partial\psi}{\partial t} = \hat{H}\psi.$$

The solution of the Schrödinger equation takes the form

$$\psi(t) = \exp(-i\hat{H}t/\hbar)\psi(0).$$

Consider as an example the Hamilton operator

$$\hat{H} = \hbar\omega\begin{pmatrix} 0 & 1 \\ 1 & 0 \end{pmatrix}.$$

Let

$$\psi(0) = \begin{pmatrix} 1 \\ 0 \end{pmatrix}$$

be the initial state. Then from the solution of the Schrödinger equation we find

$$\psi(t) = \begin{pmatrix} \cos(\omega t) & -i\sin(\omega t) \\ -i\sin(\omega t) & \cos(\omega t) \end{pmatrix}\begin{pmatrix} 1 \\ 0 \end{pmatrix} = \begin{pmatrix} \cos(\omega t) \\ -i\sin(\omega t) \end{pmatrix}.$$

Thus the probability to find the system in the initial state after a time t is given by

$$|(\psi(t), \psi(0))|^2 = \cos^2(\omega t).$$

The *Heisenberg's equation of motion* is given by

$$i\hbar\frac{dA}{dt} = [\hat{A}, \hat{H}](t).$$

Let

$$\hat{H} = \hbar\omega\sigma_z$$

be a Hamilton operator acting in the two-dimensional Hilbert space $\mathbf{C}^2$, where

$$\sigma_z := \begin{pmatrix} 1 & 0 \\ 0 & -1 \end{pmatrix}$$

and ω is the frequency. We calculate the time evolution of

$$\sigma_x := \begin{pmatrix} 0 & 1 \\ 1 & 0 \end{pmatrix}.$$

Remark. The matrices σ_x, σ_y and σ_z are the *Pauli matrices*, where

$$\sigma_y := \begin{pmatrix} 0 & -i \\ i & 0 \end{pmatrix}.$$

The Pauli matrices form a Lie algebra under the commutator.

The Heisenberg equation of motion is given by

$$i\hbar \frac{d\sigma_x}{dt} = [\sigma_x, \hat{H}](t).$$

Since

$$[\sigma_x, \hat{H}] = \hbar\omega[\sigma_x, \sigma_z] = -2i\hbar\omega\sigma_y$$

we obtain

$$\frac{d\sigma_x}{dt} = -2\omega\sigma_y(t).$$

Now we have to calculate the time-evolution of σ_y, i.e.,

$$i\hbar \frac{d\sigma_y}{dt} = [\sigma_y, \hat{H}](t).$$

Since

$$[\sigma_y, \hat{H}] = \hbar\omega[\sigma_y, \sigma_z] = 2i\hbar\omega\sigma_x$$

we find

$$\frac{d\sigma_y}{dt} = 2\omega\sigma_x(t).$$

To summarize: we have to solve the following system of linear matrix differential equations with constant coefficients

$$\frac{d\sigma_x}{dt} = -2\omega\sigma_y(t)$$

$$\frac{d\sigma_y}{dt} = 2\omega\sigma_x(t).$$

The initial conditions of this system are

$$\sigma_x(t=0) = \sigma_x, \qquad \sigma_y(t=0) = \sigma_y.$$

Then the solution of the initial value problem is given by

$$\sigma_x(t) = \sigma_x \cos(2\omega t) - \sigma_y \sin(2\omega t)$$

$$\sigma_y(t) = \sigma_y \cos(2\omega t) + \sigma_x \sin(2\omega t).$$

Remark. The solution of the Heisenberg equation of motion can also be given as

$$\sigma_x(t) = \exp(i\hat{H}t/\hbar)\sigma_x \exp(-i\hat{H}t/\hbar)$$

$$\sigma_y(t) = \exp(i\hat{H}t/\hbar)\sigma_y \exp(-i\hat{H}t/\hbar).$$

Chapter 2

Kronecker Product

2.1 Definitions and Notations

We introduce the Kronecker product of two matrices and give a number of examples.

Definition. Let A be an $m \times n$ matrix and let B be a $p \times q$ matrix. Then the *Kronecker product* of A and B is that $(mp) \times (nq)$ matrix defined by

$$A \otimes B := \begin{pmatrix} a_{11}B & a_{12}B & \cdots & a_{1n}B \\ a_{21}B & a_{22}B & \cdots & a_{2n}B \\ \vdots & & & \\ a_{m1}B & a_{m2}B & \cdots & a_{mn}B \end{pmatrix}.$$

Remark 1. Sometimes the Kronecker product is also called *direct product* or *tensor product.*

Remark 2. Instead of the definition given above some authors (Regalia and Mitra [29]) use the definition

$$A \otimes B := \begin{pmatrix} b_{11}A & b_{12}A & \cdots & b_{1q}A \\ b_{21}A & b_{22}A & \cdots & b_{2q}A \\ \vdots & & & \\ b_{p1}A & b_{p2}A & \cdots & b_{pq}A \end{pmatrix}.$$

Throughout this book we use the first definition.

Let us now give some examples.

Example. Let

$$A = \begin{pmatrix} 2 & 3 \\ 0 & 1 \end{pmatrix}, \qquad B = \begin{pmatrix} 0 & -1 \\ -1 & 1 \end{pmatrix}.$$

Then

$$A \otimes B = \begin{pmatrix} 0 & -2 & 0 & -3 \\ -2 & 2 & -3 & 3 \\ 0 & 0 & 0 & -1 \\ 0 & 0 & -1 & 1 \end{pmatrix}$$

$$B \otimes A = \begin{pmatrix} 0 & 0 & -2 & -3 \\ 0 & 0 & 0 & -1 \\ -2 & -3 & 2 & 3 \\ 0 & -1 & 0 & 1 \end{pmatrix}.$$

We see that

$$A \otimes B \neq B \otimes A. \quad \clubsuit$$

Example. Let

$$\mathbf{u} = \begin{pmatrix} 1 \\ 0 \end{pmatrix}, \qquad \mathbf{v} = \begin{pmatrix} 0 \\ 1 \end{pmatrix}.$$

Then

$$\mathbf{u} \otimes \mathbf{u} = \begin{pmatrix} 1 \\ 0 \\ 0 \\ 0 \end{pmatrix}, \qquad \mathbf{u} \otimes \mathbf{v} = \begin{pmatrix} 0 \\ 1 \\ 0 \\ 0 \end{pmatrix}$$

$$\mathbf{v} \otimes \mathbf{u} = \begin{pmatrix} 0 \\ 0 \\ 1 \\ 0 \end{pmatrix} \qquad \mathbf{v} \otimes \mathbf{v} = \begin{pmatrix} 0 \\ 0 \\ 0 \\ 1 \end{pmatrix}.$$

Now $\{\mathbf{u}, \mathbf{v}\}$ is the *standard basis* in $\mathbf{R}^2$ (or $\mathbf{C}^2$). We find that

$$\{\, \mathbf{u} \otimes \mathbf{u}, \mathbf{u} \otimes \mathbf{v}, \mathbf{v} \otimes \mathbf{u}, \mathbf{v} \otimes \mathbf{v} \,\}$$

is the standard basis in $\mathbf{R}^4$ (or $\mathbf{C}^4$). $\clubsuit$

Example. Let

$$\mathbf{u} = \frac{1}{\sqrt{2}} \begin{pmatrix} 1 \\ i \end{pmatrix}, \qquad \mathbf{v} = \frac{1}{\sqrt{2}} \begin{pmatrix} 1 \\ -i \end{pmatrix}.$$

Then

$$\mathbf{u} \otimes \mathbf{u} = \frac{1}{2} \begin{pmatrix} 1 \\ i \\ i \\ -1 \end{pmatrix}, \qquad \mathbf{u} \otimes \mathbf{v} = \frac{1}{2} \begin{pmatrix} 1 \\ -i \\ i \\ 1 \end{pmatrix}$$

$$\mathbf{v} \otimes \mathbf{u} = \frac{1}{2} \begin{pmatrix} 1 \\ i \\ -i \\ 1 \end{pmatrix}, \qquad \mathbf{v} \otimes \mathbf{v} = \frac{1}{2} \begin{pmatrix} 1 \\ -i \\ -i \\ -1 \end{pmatrix}.$$

Since $\|\mathbf{u}\| = \|\mathbf{v}\| = 1$ and $(\mathbf{u}, \mathbf{v}) = 0$, $\{\mathbf{u}, \mathbf{v}\}$ is an orthonormal basis in $\mathbf{C}^2$. We find that

$$\{\,\mathbf{u} \otimes \mathbf{u}, \mathbf{u} \otimes \mathbf{v}, \mathbf{v} \otimes \mathbf{u}, \mathbf{v} \otimes \mathbf{v}\,\}$$

is an orthonormal basis in $\mathbf{C}^4$. ♣

Example. Let

$$\mathbf{u} = \begin{pmatrix} u_1 \\ u_2 \\ u_3 \end{pmatrix}, \qquad \mathbf{v} = \begin{pmatrix} v_1 \\ v_2 \\ v_3 \end{pmatrix}.$$

Then

$$\mathbf{u}^* \equiv \bar{\mathbf{u}}^T = (\bar{u}_1, \bar{u}_2, \bar{u}_3), \qquad \mathbf{v}^* \equiv \bar{\mathbf{v}}^T = (\bar{v}_1, \bar{v}_2, \bar{v}_3)$$

and therefore

$$\mathbf{u}^* \otimes \mathbf{u} = \begin{pmatrix} \bar{u}_1 u_1 & \bar{u}_2 u_1 & \bar{u}_3 u_1 \\ \bar{u}_1 u_2 & \bar{u}_2 u_2 & \bar{u}_3 u_2 \\ \bar{u}_1 u_3 & \bar{u}_2 u_3 & \bar{u}_3 u_3 \end{pmatrix}, \qquad \mathbf{v}^* \otimes \mathbf{v} = \begin{pmatrix} \bar{v}_1 v_1 & \bar{v}_2 v_1 & \bar{v}_3 v_1 \\ \bar{v}_1 v_2 & \bar{v}_2 v_2 & \bar{v}_3 v_2 \\ \bar{v}_1 v_3 & \bar{v}_2 v_3 & \bar{v}_3 v_3 \end{pmatrix}$$

$$\mathbf{u}^* \otimes \mathbf{v} = \begin{pmatrix} \bar{u}_1 v_1 & \bar{u}_2 v_1 & \bar{u}_3 v_1 \\ \bar{u}_1 v_2 & \bar{u}_2 v_2 & \bar{u}_3 v_2 \\ \bar{u}_1 v_3 & \bar{u}_2 v_3 & \bar{u}_3 v_3 \end{pmatrix}, \qquad \mathbf{v}^* \otimes \mathbf{u} = \begin{pmatrix} \bar{v}_1 u_1 & \bar{v}_2 u_1 & \bar{v}_3 u_1 \\ \bar{v}_1 u_2 & \bar{v}_2 u_2 & \bar{v}_3 u_2 \\ \bar{v}_1 u_3 & \bar{v}_2 u_3 & \bar{v}_3 u_3 \end{pmatrix}.$$

Now

$$\mathbf{u} \otimes \mathbf{v}^* = \begin{pmatrix} u_1 \bar{v}_1 & u_1 \bar{v}_2 & u_1 \bar{v}_3 \\ u_2 \bar{v}_1 & u_2 \bar{v}_2 & u_2 \bar{v}_3 \\ u_3 \bar{v}_1 & u_3 \bar{v}_2 & u_3 \bar{v}_3 \end{pmatrix}.$$

Thus we find that

$$(\mathbf{u} \otimes \mathbf{v}^*)^* = \mathbf{u}^* \otimes \mathbf{v}.$$

Notice that

$$\mathbf{u}^* \otimes \mathbf{u} = \mathbf{u}\mathbf{u}^*. \qquad ♣$$

Example. Let

$$\mathbf{u} = \frac{1}{\sqrt{2}} \begin{pmatrix} 1 \\ 1 \end{pmatrix}, \qquad \mathbf{v} = \frac{1}{\sqrt{2}} \begin{pmatrix} 1 \\ -1 \end{pmatrix}.$$

Obviously the set $\{\,\mathbf{u}, \mathbf{v}\,\}$ forms an orthonormal basis in $\mathbf{R}^2$. Now

$$\mathbf{u}^T \otimes \mathbf{u} = \frac{1}{2} \begin{pmatrix} 1 & 1 \\ 1 & 1 \end{pmatrix}, \qquad \mathbf{v}^T \otimes \mathbf{v} = \frac{1}{2} \begin{pmatrix} 1 & -1 \\ -1 & 1 \end{pmatrix}.$$

Consequently

$$\mathbf{u}^T \otimes \mathbf{u} + \mathbf{v}^T \otimes \mathbf{v} = \begin{pmatrix} 1 & 0 \\ 0 & 1 \end{pmatrix} = I_2.$$

We notice that the 2×2 unit matrix is also given by

$$I_2 = \mathbf{u}\mathbf{u}^T + \mathbf{v}\mathbf{v}^T. \quad \clubsuit$$

Example. Let

$$\mathbf{u} = \begin{pmatrix} 1 \\ 0 \end{pmatrix}, \qquad \mathbf{v} = \begin{pmatrix} 0 \\ 1 \end{pmatrix}.$$

Then

$$\mathbf{u}^T \otimes \mathbf{u} = \begin{pmatrix} 1 & 0 \\ 0 & 0 \end{pmatrix}, \qquad \mathbf{v}^T \otimes \mathbf{v} = \begin{pmatrix} 0 & 0 \\ 0 & 1 \end{pmatrix}$$

and

$$\mathbf{u}^T \otimes \mathbf{u} + \mathbf{v}^T \otimes \mathbf{v} = \begin{pmatrix} 1 & 0 \\ 0 & 1 \end{pmatrix} = I_2.$$

We notice that the 2×2 unit matrix is also given by

$$I_2 = \mathbf{u}\mathbf{u}^T + \mathbf{v}\mathbf{v}^T. \quad \clubsuit$$

Remark. The last two examples indicate that the unit matrix can be represented with the help of an orthonormal basis in $\mathbf{R}^2$ and the Kronecker product. We address this problem in section 2.8 in more detail.

Example. Let I_n be the $n \times n$ unit matrix and let I_m be the $m \times m$ unit matrix. Then $I_n \otimes I_m$ is the $(nm) \times (mn)$ unit matrix. Obviously,

$$I_n \otimes I_m = I_m \otimes I_n = I_{m \times n}. \quad \clubsuit$$

Example. Let A_n be an arbitrary $n \times n$ matrix and let O_m be the $m \times m$ zero matrix. Then

$$A_n \otimes O_m = O_{m \times n}. \quad \clubsuit$$

Exercises. (1) Let A and B be $n \times n$ diagonal matrices. Show that $A \otimes B \neq B \otimes A$ in general. What is the condition on A and B such that $A \otimes B = B \otimes A$?

(2) Let A, B be 2×2 matrices. Find the conditions on A and B such that

$$A \otimes B = B \otimes A .$$

(3) Let

$$X = \begin{pmatrix} 16 & 3 & 2 & 13 \\ 5 & 10 & 11 & 8 \\ 9 & 6 & 7 & 12 \\ 4 & 15 & 14 & 1 \end{pmatrix} .$$

X is a so-called *magic square*, since its row sums, column sum, prinicpal diagonal sum, and principal counterdiagonal are equal. Is $X \otimes X$ a magic square ?

(4) A *Toeplitz matrix* is a square matrix whose elements are the same along any northwest (NW) to southeast (SE) diagonal. The Toeplitz matrix of size $2^2 \times 2^2$ is designated $T(2)$ and given by

$$T(2) = \begin{pmatrix} a_{11} & a_{12} & a_{13} & a_{14} \\ a_{21} & a_{11} & a_{12} & a_{13} \\ a_{31} & a_{21} & a_{11} & a_{12} \\ a_{41} & a_{31} & a_{21} & a_{11} \end{pmatrix} .$$

Is $T(2) \otimes T(2)$ a Toeplitz matrix ?

(5) A *circulant matrix* is a square matrix whose elements in each row are obtained by a circular right shift of the elements in the preceding row. An example is

$$C(2) = \begin{pmatrix} c_{11} & c_{12} & c_{13} & c_{14} \\ c_{14} & c_{11} & c_{12} & c_{13} \\ c_{13} & c_{14} & c_{11} & c_{12} \\ c_{12} & c_{13} & c_{14} & c_{11} \end{pmatrix} .$$

Given the elements of any row, the entire matrix can be developed. Is $C(2) \otimes C(2)$ a circulant matrix ?

(6) A vector $\mathbf{u}^T = (u_1, u_2, \ldots, u_n)$ is called a *probability vector* if the components are nonnegative and their sum is 1. Let $\mathbf{u}^T$ and $\mathbf{v}^T$ be probability vectors. Show that $\mathbf{u}^T \otimes \mathbf{v}^T$ is a probability vector.

(7) A square matrix $A = (a_{jk})$ is called a *stochastic matrix* if each of its rows is a probability vector, i.e. if each entry of A is nonnegative and the sum of the entries in each row is equal to 1. Let A and B be $n \times n$ stochastic matrices. Show that AB is a stochastic matrix. Show that $A \otimes B$ is a stochastic matrix.

(8) Let A, B and C be 3×3 matrices. We set

$$D := \begin{pmatrix} A & B & 0 \\ C & A & B \\ 0 & C & A \end{pmatrix}.$$

Show that D can be written in the form

$$D = (I_3 \otimes A + L \otimes C + U \otimes B)$$

where L and U are defined as

$$L = \begin{pmatrix} 0 & 0 & 0 \\ 1 & 0 & 0 \\ 0 & 1 & 0 \end{pmatrix}, \qquad U = \begin{pmatrix} 0 & 1 & 0 \\ 0 & 0 & 1 \\ 0 & 0 & 0 \end{pmatrix}.$$

(9) The symmmetric matrix of order n

$$H = (h_{ij}), \qquad h_{ij} = \frac{1}{i+j-1}, \quad i,j = 1,2,\ldots,n$$

is called a *Hilbert matrix*. Show that this matrix is positive definite. The number $\text{cond}_2(H)$ is defined as

$$\text{cond}_2(H) := \frac{\max_i |\lambda_i(H)|}{\min_i |\lambda_i(H)|}.$$

Find $\text{cond}_2(H)$ for $n = 2, 3, 4$, and 5. Show that

$$H \otimes H$$

is not a Hilbert matrix.

(10) Prove the identity

$$(A + \beta \mathbf{u} \otimes \mathbf{v}^T)^{-1} \equiv A^{-1} - \frac{\beta}{1 + \beta \mathbf{v}^T A^{-1} \mathbf{u}} A^{-1} \mathbf{u} \otimes \mathbf{v}^T A^{-1}$$

where A is an invertible matrix of order n, $\beta \in \mathbf{R}$ and $\mathbf{u}, \mathbf{v} \in \mathbf{R}^n$.

Remark. This identity is sometimes used to find the inverse of a matrix which appears in the form of a perturbation $A + \beta \mathbf{u} \otimes \mathbf{v}^T$ of a matrix A whose inverse is known.

(11) Let $\mathbf{u}$, $\mathbf{v} \in \mathbf{R}^{\mathbf{n}}$. Assume that $||\mathbf{u}|| = 1$ and $||\mathbf{v}|| = 1$. Is $||\mathbf{u} \otimes \mathbf{v}|| = 1$?

2.2 Basic Properties

Here we list some basic properties of the Kronecker product. These properties have been listed in various books on linear algebra and matrix theory ([15], [21], [30]). In almost all cases the proof is straightforward and is left to the reader as an exercise.

Let A be an $m \times n$ matrix, B be a $p \times q$ matrix and C be an $s \times t$ matrix. Then

$$(A \otimes B) \otimes C \equiv A \otimes (B \otimes C).$$

The proof is straightforward. The matrix $A \otimes B \otimes C$ has (mps) rows and (nqt) columns. This means the Kronecker product satisfies the *associative law.*

Let A be an $m \times n$ matrix and B be a $p \times q$ matrix. Let $c \in \mathbf{C}$. Then

$$(cA) \otimes B \equiv c(A \otimes B) \equiv A \otimes (cB).$$

Let A and B be $m \times n$ matrices and C and D be $p \times q$ matrices. Then we have

$$(A + B) \otimes (C + D) \equiv A \otimes C + A \otimes D + B \otimes C + B \otimes D.$$

This means the Kronecker product obeys the *distributive law.*

We can easily prove that the row rank and the column rank are equal to the same number. Let $r(A)$ be the rank of A and $r(B)$ be the rank of B. Then

$$r(A \otimes B) = r(A)r(B).$$

Example. Let

$$A = \begin{pmatrix} 0 & 1 \\ 1 & 0 \end{pmatrix}, \qquad B = \begin{pmatrix} 0 & 1 \\ 0 & 0 \end{pmatrix}.$$

Then $r(A) = 2$ and $r(B) = 1$. Therefore $r(A \otimes B) = 2$. ♣

Let A be an $m \times n$ matrix and B be an $p \times q$ matrix. The following properties can be easily proved

$$\begin{aligned}(A \otimes B)^T &\equiv A^T \otimes B^T \\ \overline{(A \otimes B)} &\equiv \bar{A} \otimes \bar{B} \\ (A \otimes B)^* &\equiv A^* \otimes B^*.\end{aligned}$$

Given matrices of a special type we now summarize for which matrices the Kronecker product is also of this type.

(1) If A and B are diagonal matrices, then $A \otimes B$ is a diagonal matrix. Conversely, if $A \otimes B$ is a *diagonal matrix* and $A \otimes B \neq 0$, then A and B are diagonal matrices.

(2) Let A and B be upper triangular matrices, then $A \otimes B$ is an upper *triangular matrix*. Similarly, if A and B are lower triangular, then $A \otimes B$ is a lower triangular matrix.

(3) Let A and B be Hermitian matrices. Then $A \otimes B$ is a *Hermitian matrix.*

(4) Let A and B be normal matrices. Then $A \otimes B$ is a *normal matrix.*

(5) Let A and B be Hermitian matrices. Let A and B be *positive definite.* Then $A \otimes B$ is a positive definite matrix. The same holds for positive semi-definite matrices.

(6) Let A be an *invertible* $n \times n$ matrix. Let B be an invertible $m \times m$ matrix. Then $A \otimes B$ is an invertible matrix

(7) Let U and V be *unitary* $n \times n$ matrices. Then $U \otimes V$ is a unitary matrix. The proof will be given in section 2.3.

Definition. The *Kronecker powers* of an $m \times n$ matrix A are defined as

$$A^{[2]} := A \otimes A$$

and, in general

$$A^{[k+1]} := A \otimes A^{[k]}$$

where $k \in \mathbf{N}$. We find the following property

$$A^{[k+l]} = A^{[k]} \otimes A^{[l]}.$$

Example. In genetics we find the following example (Searle [30]). Consider the n generations of random mating starting with the progeny obtained from crossing two autotetraploid plants which both have genotype AAaa. Normally the original plants would produce gametes AA, Aa and aa in the proportion 1:4:1. However suppose the proportion is

$$u : 1 - 2u : u$$

where, for example, u might take the value $(1-\alpha)/6$, for α being a measure of "diploidization" of plants: $\alpha = 0$ is the case of autotetraploids with chromosome segregation and $\alpha = 1$ is the diploid case with all gametes being Aa. What are the genotypic frequencies in the population after n generations of random mating ? Let $\mathbf{u}_j$ be the vector of gametic frequencies and $\mathbf{f}_j$ the vector of genotype frequencies in the j-th generation of random mating, where $\mathbf{u}_0$ is the vector of gametic frequencies in the initial plants. Then

$$\mathbf{u}_0 = \begin{pmatrix} u \\ 1-2u \\ u \end{pmatrix}$$

and

$$\mathbf{f}_{j+1} = \mathbf{u}_j \otimes \mathbf{u}_j$$

for $j = 0, 1, 2, \ldots, n$. Furthermore, the relationship between $\mathbf{u}_j$ and $\mathbf{f}_j$ at any generation is given by

$$\mathbf{u}_j = B\mathbf{f}_j$$

where

$$B = \begin{pmatrix} 1 & \frac{1}{2} & u & \frac{1}{2} & u & 0 & u & 0 & 0 \\ 0 & \frac{1}{2} & 1-2u & \frac{1}{2} & 1-2u & \frac{1}{2} & 1-2u & \frac{1}{2} & 0 \\ 0 & 0 & u & 0 & u & \frac{1}{2} & u & \frac{1}{2} & 1 \end{pmatrix}.$$

Thus

$$\begin{aligned} \mathbf{f}_j &= \mathbf{u}_{j-1} \otimes \mathbf{u}_{j-1} \\ &= B\mathbf{f}_{j-1} \otimes B\mathbf{f}_{j-1} \\ &= (B \otimes B)(\mathbf{f}_{j-1} \otimes \mathbf{f}_{j-1}) \\ &= (B \otimes B)[(B \otimes B)(\mathbf{f}_{j-2} \otimes \mathbf{f}_{j-2}) \otimes (B \otimes B)(\mathbf{f}_{j-2} \otimes \mathbf{f}_{j-2})] \\ &= (B \otimes B)[(B \otimes B) \otimes (B \otimes B)][(\mathbf{f}_{j-2} \otimes \mathbf{f}_{j-2}) \otimes (\mathbf{f}_{j-2} \otimes \mathbf{f}_{j-2})]. \end{aligned}$$

It can be verified by induction that

$$\mathbf{f}_j = \otimes^2 B(\otimes^4 B)(\otimes^8 B) \cdots (\otimes^{2^{j-1}} B)(\otimes^{2^j} \mathbf{u}_0)$$

where $\otimes^n B$ denotes the Kronecker product of n B's. ♣

Exercises. (1) Let

$$A = \begin{pmatrix} 2 & 0 & 1 \\ 3 & 1 & 0 \end{pmatrix}, \qquad B = \begin{pmatrix} 1 & 2 & 2 \\ 1 & 2 & 0 \end{pmatrix}.$$

Find $r(A)$, $r(B)$ and $r(A \otimes B)$.

(2) Let A be an $n \times n$ matrix and I be the $n \times n$ unit matrix. Show that in general $I \otimes A \neq A \otimes I$.

(3) An $n \times n$ matrix A is called *skew-Hermitian* if $A^* = -A$. Let A and B be $n \times n$ skew-Hermitian matrices. Is the matrix $A \otimes B$ skew-Hermitian ?

(4) Show that the Kronecker product of two positive-definite matrices is again a positive-definite matrix.

(5) Show that the Kronecker product of two Hermitian matrices is again a Hermitian matrix.

(6) An $n \times n$ matrix A is called *nilpotent* if

$$A^k = 0$$

for some positive integer k. Let A and B be two $n \times n$ nilpotent matrices. Show that $A \otimes B$ is nilpotent.

(7) An $n \times n$ matrix A is called *idempotent* if

$$A^k = A$$

for some positive integer k. Let A and B be two $n \times n$ idempotent matrices. Show that $A \otimes B$ is idempotent.

(8) The *elementary matrix* (E_{ij}) is defined as the matrix of order $m \times n$ which has 1 in the (i, j)-th position and all other elements are zero. Let (E_{ij}) and (F_{kl}) be elementary matrices of order $m \times n$ and $p \times q$, respectively. Is

$$(E_{ij}) \otimes (F_{kl})$$

an elementary matrix ?

2.3 Matrix Multiplication

In this section we describe the connection of the matrix multiplication and the Kronecker product (Gröbner [15], Lancaster [21], Brewer [8], Davis [10], Searle [30]).

Let A be an $m \times n$ matrix and B be an $n \times r$ matrix. Then the *matrix product* AB is defined and the matrix elements of AB are given by

$$(AB)_{kl} := \sum_{j=1}^{n} a_{kj} b_{jl}.$$

If A is of order $m \times n$, if B is of order $n \times p$, and if C is of order $p \times q$, then

$$A(BC) = (AB)C.$$

Now we can discuss the connection of Kronecker products and matrix products. Assume that the matrix A is of order $m \times n$, B of order $p \times q$, C of order $n \times r$ and D of order $q \times s$. Then by straightforward calculation we can prove that

$$(A \otimes B)(C \otimes D) = (AC) \otimes (BD).$$

Example. Let A and B be $n \times n$ matrices. Then

$$(A \otimes I_n)(I_n \otimes B) = A \otimes B. \quad \clubsuit$$

Now we can prove the following theorems.

Theorem. Let A be an invertible $m \times m$ matrix. Let B be an invertible $n \times n$ matrix. Then $A \otimes B$ is invertible and

$$(A \otimes B)^{-1} = A^{-1} \otimes B^{-1}.$$

Proof. Since $\det A \neq 0$, $\det B \neq 0$ and

$$\det(A \otimes B) = (\det A)^n (\det B)^m$$

we have

$$\det(A \otimes B) \neq 0.$$

The proof of the identity $\det(A \otimes B) = (\det A)^n (\det B)^m$ will be given in section (2.5). Thus $(A \otimes B)^{-1}$ exists and

$$(A \otimes B)^{-1}(A \otimes B) = I_{mn}$$

or

$$(A^{-1} \otimes B^{-1})(A \otimes B) = (A^{-1}A) \otimes (B^{-1}B) = I_{n^2}. \quad \spadesuit$$

Theorem. Let U and V be unitary matrices. Then $U \otimes V$ is a unitary matrix.

Proof. Since U and V are unitary matrices we have

$$U^* = U^{-1}, \qquad V^* = V^{-1}.$$

Therefore

$$(U \otimes V)^* = U^* \otimes V^* = U^{-1} \otimes V^{-1} = (U \otimes V)^{-1}. \quad \spadesuit$$

The equation $(A \otimes B)(C \otimes D) = (AC) \otimes (BD)$ can be extended as follows

$$(A_1 \otimes B_1)(A_2 \otimes B_2)(A_3 \otimes B_3) = (A_1 A_2 A_3) \otimes (B_1 B_2 B_3)$$

and

$$(A_1 \otimes A_2 \otimes A_3)(B_1 \otimes B_2 \otimes B_3) = (A_1 B_1) \otimes (A_2 B_2) \otimes (A_3 B_3)$$

where it is assumed that the matrix products exist. The extension to r-factors is

$$(A_1 \otimes B_1)(A_2 \otimes B_2) \cdots (A_r \otimes B_r) = (A_1 A_2 \cdots A_r) \otimes (B_1 B_2 \cdots B_r)$$

and

$$(A_1 \otimes A_2 \otimes \cdots \otimes A_r)(B_1 \otimes B_2 \otimes \cdots \otimes B_r) = (A_1 B_1) \otimes (A_2 B_2) \otimes \cdots \otimes (A_r B_r).$$

Let us now study special cases of the identity

$$(A \otimes B)(C \otimes D) = (AC) \otimes (BD).$$

If $A = I_n$, then the identity takes the form

$$(I_n \otimes B)(C \otimes D) = C \otimes (BD).$$

If $D = I_q$, then we find

$$(A \otimes B)(C \otimes I_q) = (AC) \otimes B.$$

Let A be an $m \times m$ matrix and B be an $n \times n$ matrix. Then

$$A \otimes B = (A \otimes I_n)(I_m \otimes B) = (I_m \otimes B)(A \otimes I_n).$$

Let

$$f(x) := \sum_{j=0}^{r} a_j x^j \equiv a_0 + a_1 x + a_2 x^2 + \cdots + a_r x^r$$

be a *polynomial*. Let A be an $n \times n$ matrix. Then

$$f(A) = \sum_{j=0}^{r} a_j A^j \equiv a_0 I_n + a_1 A + a_2 A^2 + \cdots + a_r A^r$$

where I_n is the $n \times n$ unit matrix. Now we can easily prove that

$$f(I_n \otimes A) \equiv I_n \otimes f(A)$$

and

$$f(A \otimes I_n) \equiv f(A) \otimes I_n.$$

To prove these identities one applies the equation $(A \otimes B)(C \otimes D) = (AC) \otimes (BD)$ repeatetly.

Definition. Two $n \times n$ matrices A and B are called *similar* if there exists an invertible $n \times n$ matrix Q such that

$$B = Q^{-1}AQ.$$

Theorem. Let A and B be similar $n \times n$ matrices. Then $A \otimes A$ and $B \otimes B$ are similar and $A \otimes B$ and $B \otimes A$ are similar.

The proof is straightforward and left to the reader.

Theorem. Let A and B be $m \times m$ matrices and C, D be $n \times n$ matrices. Assume that

$$[A, B] = 0, \qquad [C, D] = 0$$

where

$$[A, B] := AB - BA$$

defines the *commutator*. Then

$$[A \otimes C, B \otimes D] = 0.$$

Proof.

$$\begin{aligned} [A \otimes C, B \otimes D] &= (A \otimes C)(B \otimes D) - (B \otimes D)(A \otimes C) \\ &= (AB) \otimes (CD) - (BA) \otimes (DC) \\ &= (AB) \otimes (CD) - (AB) \otimes (CD) \\ &= 0 \end{aligned}$$

where we have used $(A \otimes B)(C \otimes D) = (AC) \otimes (BD)$. ♠

From this theorem we find the corollary

Corollary. Let A be an $m \times m$ matrix and B be an $n \times n$ matrix. I_m and I_n denote the $m \times m$ and $n \times n$ unit matrix, respectively. Then

$$[A \otimes I_n, I_m \otimes B] = 0$$

where 0 is the $(mm) \times (nn)$ zero matrix.

Proof.

$$\begin{aligned}
[A \otimes I_n, I_m \otimes B] &= (A \otimes I_n)(I_m \otimes B) - (I_m \otimes B)(A \otimes I_n) \\
&= A \otimes B - A \otimes B \\
&= 0. \quad \spadesuit
\end{aligned}$$

Using these results we are able to prove the following theorem

Theorem. Let A and B be $n \times n$ matrices. I denotes the $n \times n$ unit matrix. Then

$$\exp(A \otimes I + I \otimes B) \equiv (\exp A) \otimes (\exp B).$$

Proof. Since

$$[A \otimes I, I \otimes B] = 0$$

we have

$$\exp(A \otimes I + I \otimes B) = \exp(A \otimes I) \exp(I \otimes B).$$

Now

$$\exp(A \otimes I) := \sum_{k=0}^{\infty} \frac{(A \otimes I)^k}{k!}$$

and

$$\exp(I \otimes B) := \sum_{k=0}^{\infty} \frac{(I \otimes B)^k}{k!}.$$

An arbitrary term in the expansion of $\exp(A \otimes I) \exp(I \otimes B)$ is given by

$$\frac{1}{n!} \frac{1}{m!} (A \otimes I)^n (I \otimes B)^m.$$

Since

$$(A \otimes I)^n (I \otimes B)^m \equiv (A^n \otimes I^n)(I^m \otimes B^m) \equiv (A^n \otimes I)(I \otimes B^m) \equiv (A^n \otimes B^m)$$

we obtain

$$\frac{1}{n!} \frac{1}{m!} (A \otimes I)^n (I \otimes B)^m \equiv \left(\frac{1}{n!} A^n\right) \otimes \left(\frac{1}{m!} B^m\right).$$

This proves the theorem. $\spadesuit$

An extension of the formula given above is

Theorem. Let $A_1, A_2, \ldots, A_r$ be real $n \times n$ matrices and let I be the $n \times n$ identity matrix. Then

$$\exp(A_1 \otimes I \otimes \ldots \otimes I + I \otimes A_2 \otimes \ldots \otimes I + \cdots + I \otimes \cdots \otimes I \otimes A_r)$$

$$\equiv \exp(A_1) \otimes \exp(A_2) \otimes \cdots \otimes \exp(A_r).$$

The proof is left as an exercise for the reader.

We can consider also other analytic functions. For example, we have

$$\cos(I_n \otimes A) \equiv I_n \otimes \cos(A).$$

To prove this identity one uses the expansion

$$\cos(x) := \sum_{k=0}^{\infty} \frac{(-1)^k x^{2k}}{(2k)!}.$$

Example. Let A be an $m \times m$ matrix and B be an $n \times n$ matrix. We have

$$\sin(A \otimes I_n + I_m \otimes B^T) \equiv \sin(A) \otimes \cos(B^T) + \cos(A) \otimes \sin(B^T). \quad \clubsuit$$

Definition. Let X and Y be $n \times n$ matrices. Then the *anticommutator* $[X, Y]_+$ of X and Y is defined as

$$[X, Y]_+ := XY + YX.$$

The anti-commutator plays an important role for Fermi particles (see chapter 3).

Example. Let

$$\sigma_z := \begin{pmatrix} 1 & 0 \\ 0 & -1 \end{pmatrix}, \qquad \sigma_+ := \begin{pmatrix} 0 & 2 \\ 0 & 0 \end{pmatrix}$$

and let I be the 2×2 unit matrix. We set

$$A := \sigma_+ \otimes I, \qquad B := \sigma_z \otimes \sigma_+ .$$

Since

$$\sigma_+ \sigma_z = \begin{pmatrix} 0 & -2 \\ 0 & 0 \end{pmatrix}, \qquad \sigma_z \sigma_+ = \begin{pmatrix} 0 & 2 \\ 0 & 0 \end{pmatrix}$$

we find that

$$[A, B]_+ = 0.$$

Exercises. (1) Let $\mathbf{u}, \mathbf{v} \in \mathbf{C}^n$. Show that

$$\mathbf{u}^T \otimes \mathbf{v} = \mathbf{v}\mathbf{u}^T = \mathbf{v} \otimes \mathbf{u}^T.$$

(2) Let A be a Hermitian matrix. Show that

$$(I + iA)^{-1}(I - iA)$$

is unitary. Is the matrix

$$(I + iA)^{-1} \otimes (I - iA)$$

unitary ?

(3) Let $\mathbf{u}$ and $\mathbf{v}$ be column vectors in $\mathbf{C}^n$. Let A and B be $n \times n$ matrices over $\mathbf{C}$. Show that

$$(\mathbf{u}^T \otimes B)(A \otimes \mathbf{v}) = (B\mathbf{v})(\mathbf{u}^T A).$$

(4) Let A be an $n \times n$ matrix over $\mathbf{C}$. Let $\mathbf{u}$ be a column vector in $\mathbf{C}^n$. Show that

$$(A \otimes A)(\mathbf{u} \otimes \mathbf{u}) = (A \otimes (A\mathbf{u}))\mathbf{u}.$$

(5) Let A be an $n \times n$ matrix and $\mathbf{u} \in \mathbf{C}^n$. Show that

$$(I_n \otimes \mathbf{u})A = A \otimes \mathbf{u}.$$

(6) Let A be an $n \times n$ matrix. Let $\mathbf{u}, \mathbf{v} \in \mathbf{C}^n$. Show that

$$A(\mathbf{u} \otimes \mathbf{v}^T)A = (A\mathbf{u}) \otimes (\mathbf{v}^T A).$$

(7) Let A, B be $n \times n$ matrices. Show that

$$(A + B)^{-1} = (I + A^{-1}B)^{-1}A^{-1}$$

where it is assumed that the inverse matrices exist. Find $(A + B)^{-1} \otimes (A + B)^{-1}$.

(8) Let A be an $n \times n$ matrix. Assume that A^{-1} exists. Let $\mathbf{u}, \mathbf{v} \in \mathbf{C}^n$. We define

$$\lambda := \mathbf{v}^T A^{-1}\mathbf{u}$$

and assume that $|\lambda| < 1$. Show that

$$(A + \mathbf{u} \otimes \mathbf{v}^T)^{-1} = A^{-1} - \frac{(A^{-1}\mathbf{u}) \otimes (\mathbf{v}^T A^{-1})}{1 + \lambda}.$$

2.4 Permutation Matrices

Permutation matrices have been introduced in section 1.7. In this section we describe the connection with the Kronecker product.

Theorem. Let P and Q be permutation matrices of order n and m, respectively. Then $P \otimes Q$ and $Q \otimes P$ are permutation matrices of order nm.

The proof is straigthforward and is left to the reader.

Example. Let

$$P = \begin{pmatrix} 0 & 1 \\ 1 & 0 \end{pmatrix}, \qquad Q = \begin{pmatrix} 0 & 1 & 0 \\ 1 & 0 & 0 \\ 0 & 0 & 1 \end{pmatrix}$$

be two permutation matrices. Then

$$P \otimes Q = \begin{pmatrix} 0 & 0 & 0 & 0 & 1 & 0 \\ 0 & 0 & 0 & 1 & 0 & 0 \\ 0 & 0 & 0 & 0 & 0 & 1 \\ 0 & 1 & 0 & 0 & 0 & 0 \\ 1 & 0 & 0 & 0 & 0 & 0 \\ 0 & 0 & 1 & 0 & 0 & 0 \end{pmatrix}, \qquad Q \otimes P = \begin{pmatrix} 0 & 0 & 0 & 1 & 0 & 0 \\ 0 & 0 & 1 & 0 & 0 & 0 \\ 0 & 1 & 0 & 0 & 0 & 0 \\ 1 & 0 & 0 & 0 & 0 & 0 \\ 0 & 0 & 0 & 0 & 0 & 1 \\ 0 & 0 & 0 & 0 & 1 & 0 \end{pmatrix}$$

are permutation matrices. We see that

$$P \otimes Q \neq Q \otimes P. \quad \clubsuit$$

Theorem. Let A be an $m \times n$ matrix and B be a $p \times q$ matrix, respectively. Then there exist permutation matrices P and Q such that

$$B \otimes A = P(A \otimes B)Q.$$

The proof is left as an exercise to the reader.

Remark. The Kronecker product $B \otimes A$ contains the same entries as $A \otimes B$ only in another order.

Example. Let

$$A = \begin{pmatrix} a_{11} & a_{12} \\ a_{21} & a_{22} \end{pmatrix}, \qquad B = \begin{pmatrix} b_{11} & b_{12} \\ b_{21} & b_{22} \end{pmatrix}.$$

Then

$$A \otimes B = \begin{pmatrix} a_{11}b_{11} & a_{11}b_{12} & a_{12}b_{11} & a_{12}b_{12} \\ a_{11}b_{21} & a_{11}b_{22} & a_{12}b_{21} & a_{12}b_{22} \\ a_{21}b_{11} & a_{21}b_{12} & a_{22}b_{11} & a_{22}b_{12} \\ a_{21}b_{21} & a_{21}b_{22} & a_{22}b_{21} & a_{22}b_{22} \end{pmatrix}.$$

and

$$B \otimes A = \begin{pmatrix} b_{11}a_{11} & b_{11}a_{12} & b_{12}a_{11} & b_{12}a_{12} \\ b_{11}a_{21} & b_{11}a_{22} & b_{12}a_{21} & b_{12}a_{22} \\ b_{21}a_{11} & b_{21}a_{12} & b_{22}a_{11} & b_{22}a_{12} \\ b_{21}a_{21} & b_{21}a_{22} & b_{22}a_{21} & b_{22}a_{22} \end{pmatrix}.$$

Let

$$P = Q = \begin{pmatrix} 1 & 0 & 0 & 0 \\ 0 & 0 & 1 & 0 \\ 0 & 1 & 0 & 0 \\ 0 & 0 & 0 & 1 \end{pmatrix}.$$

Then

$$B \otimes A = P(A \otimes B)Q. \quad \clubsuit$$

We have already mentioned that the $n \times n$ permutation matrices form a finite group and the number of group elements is given by $n! \equiv 1 \cdot 2 \cdot 3 \cdot \ldots \cdot n$. The set of all 2×2 permutation matrices are given by

$$\left\{ I = \begin{pmatrix} 1 & 0 \\ 0 & 1 \end{pmatrix}, \quad P = \begin{pmatrix} 0 & 1 \\ 1 & 0 \end{pmatrix} \right\}$$

with $II = I$, $IP = P$, $PI = P$, $PP = I$. Then we obtain

$$I \otimes I = \begin{pmatrix} 1 & 0 & 0 & 0 \\ 0 & 1 & 0 & 0 \\ 0 & 0 & 1 & 0 \\ 0 & 0 & 0 & 1 \end{pmatrix}, \qquad I \otimes P = \begin{pmatrix} 0 & 1 & 0 & 0 \\ 1 & 0 & 0 & 0 \\ 0 & 0 & 0 & 1 \\ 0 & 0 & 1 & 0 \end{pmatrix}$$

$$P \otimes I = \begin{pmatrix} 0 & 0 & 1 & 0 \\ 0 & 0 & 0 & 1 \\ 1 & 0 & 0 & 0 \\ 0 & 1 & 0 & 0 \end{pmatrix}, \qquad P \otimes P = \begin{pmatrix} 0 & 0 & 0 & 1 \\ 0 & 0 & 1 & 0 \\ 0 & 1 & 0 & 0 \\ 1 & 0 & 0 & 0 \end{pmatrix}.$$

Since

$$\begin{aligned}
(I \otimes I)(I \otimes I) &= I \otimes I \\
(I \otimes I)(I \otimes P) &= (I \otimes P)(I \otimes I) = I \otimes P \\
(I \otimes I)(P \otimes I) &= (P \otimes I)(I \otimes I) = P \otimes I \\
(P \otimes P)(I \otimes I) &= (I \otimes I)(P \otimes P) = P \otimes P \\
(P \otimes P)(I \otimes P) &= (I \otimes P)(P \otimes P) = P \otimes I \\
(P \otimes I)(P \otimes P) &= (P \otimes P)(P \otimes I) = I \otimes P \\
(P \otimes P)(P \otimes P) &= I \otimes I
\end{aligned}$$

we find that the set $\{ I\otimes I, I\otimes P, P\otimes I, P\otimes P \}$ forms a group under matrix multiplication. It is a subgroup of the group of the 4×4 permutation matrices. This finite group consists of 24 matrices.

Exercises. (1) Let

$$\{ P_1, \ldots, P_r \}$$

be the set of all $n \times n$ permutation matrices, i.e. $r = n!$. Show that

$$\{ P_1 \otimes P_1, \ldots, P_1 \otimes P_r, P_2 \otimes P_1, \ldots, P_r \otimes P_r \}$$

are permutation matrices and form a group under matrix multiplication. Show that they form a subgroup of the $(n \cdot n) \times (n \cdot n)$ permutation matrices.

(2) Let P and Q be permutation matrices. Is $(P \otimes Q)(Q \otimes P)$ a permutation matrix ? Is $(P \otimes P)(Q \otimes Q)$ a permutation matrix ?

(3) Let P be an $n \times n$ permutation matrix with $\mathrm{tr}(P) = k$, where $k \in \mathbf{N}_0$. Let Q be an $m \times m$ with $\mathrm{tr}(Q) = l$, where $l \in \mathbf{N}_0$. Find $\mathrm{tr}(P \otimes Q)$.

(4) Let P be an $n \times n$ and Π be an $m \times m$ projection matrix. Is $P \otimes \Pi$ a projection matrix ? Is $P \otimes \Pi \otimes P^{-1}$ a projection matrix ?

(5) Let

$$P = \begin{pmatrix} 0 & 1 & 0 \\ 0 & 0 & 1 \\ 1 & 0 & 0 \end{pmatrix}.$$

Find P^2, P^3, $P \otimes P$, $(P \otimes P)^2$, and $(P \otimes P)^3$. Discuss the result. Show that the permutation matrix P can be written as

$$P = M_1 + \chi^2 M_2 + \chi M_3$$

where

$$\chi := \exp(2\pi i/3)$$

and

$$M_1 := \frac{1}{3}\begin{pmatrix} 1 & 1 & 1 \\ 1 & 1 & 1 \\ 1 & 1 & 1 \end{pmatrix}, \quad M_2 := \frac{1}{3}\begin{pmatrix} 1 & \chi & \chi^2 \\ \chi^2 & 1 & \chi \\ \chi & \chi^2 & 1 \end{pmatrix}, \quad M_3 := \frac{1}{3}\begin{pmatrix} 1 & \chi^2 & \chi \\ \chi & 1 & \chi^2 \\ \chi^2 & \chi & 1 \end{pmatrix}.$$

Show that

$$M_1^2 = M_1, \qquad M_2^2 = M_2, \qquad M_3^2 = M_3 .$$

2.5 Trace and Determinant

In this section we describe the connection of the Kronecker product and trace and determinate of square matrices.

Let A be an $m \times m$ matrix and B be an $n \times n$ matrix. Then the matrix $A \otimes B$ is an $(mn) \times (mn)$ matrix. So we can calculate $\mathrm{tr}(A \otimes B)$ and we find the following theorem

Theorem. Let A be an $m \times m$ matrix and B be an $n \times n$ matrix. Then

$$\mathrm{tr}(A \otimes B) \equiv (\mathrm{tr}A)(\mathrm{tr}B).$$

Proof. From the definition of the Kronecker product we find

$$\mathrm{diag}(a_{jj}B) = (a_{jj}b_{11}, a_{jj}b_{22}, \ldots, a_{jj}b_{nn}).$$

Therefore

$$\mathrm{tr}(A \otimes B) = \sum_{j=1}^{m} \sum_{k=1}^{n} a_{jj} b_{kk}.$$

Since

$$\mathrm{tr}A = \sum_{j=1}^{m} a_{jj}, \qquad \mathrm{tr}B = \sum_{k=1}^{n} b_{kk}$$

we find the identity. ♠

In statistical mechanics the following theorem plays an important role.

Theorem. Let A, B be two arbitrary $n \times n$ matrixes. Let I be the $n \times n$ unit matrix. Then

$$\mathrm{tr}\exp(A \otimes I + I \otimes B) \equiv (\mathrm{tr}\exp(A))(\mathrm{tr}\exp(B)).$$

Proof. Since

$$[A \otimes I, I \otimes B] = 0$$

we have

$$\mathrm{tr}\exp(A \otimes I + I \otimes B) = \mathrm{tr}\left(\exp(A \otimes I)\exp(I \otimes B)\right).$$

Now

$$\exp(A \otimes I) := \sum_{k=0}^{\infty} \frac{(A \otimes I)^k}{k!}$$

and

$$\exp(I \otimes B) := \sum_{k=0}^{\infty} \frac{(I \otimes B)^k}{k!}.$$

An arbitrary term in the expansion of $\exp(A \otimes I)\exp(I \otimes B)$ is given by

$$\frac{1}{n!}\frac{1}{m!}(A \otimes I)^n (I \otimes B)^m.$$

Now we have

$$(A \otimes I)^n (I \otimes B)^m \equiv (A^n \otimes I^n)(I^m \otimes B^m) \equiv (A^n \otimes I)(I \otimes B^m) \equiv (A^n \otimes B^m).$$

Therefore

$$\mathrm{tr}\left(\frac{1}{n!}\cdot\frac{1}{m!}(A \otimes I)^n (I \otimes B)^m\right) \equiv \frac{1}{n!}\frac{1}{m!}(\mathrm{tr}A^n)(\mathrm{tr}B^m).$$

Since

$$(\mathrm{tr}\exp(A))(\mathrm{tr}\exp(B)) \equiv \mathrm{tr}\left(\sum_{k=0}^{\infty}\frac{A^k}{k!}\right)\mathrm{tr}\left(\sum_{k=0}^{\infty}\frac{B^k}{k!}\right) \equiv \left(\sum_{k=0}^{\infty}\frac{1}{k!}\mathrm{tr}A^k\right)\left(\sum_{k=0}^{\infty}\frac{1}{k!}\mathrm{tr}B^k\right)$$

we have proved the identity. ♠

An extension of the above formula is

$$\mathrm{tr}\exp(A_1 \otimes I \otimes \ldots \otimes I + I \otimes A_2 \otimes \ldots \otimes I + \cdots + I \otimes \cdots \otimes I \otimes A_r)$$
$$\equiv (\mathrm{tr}\exp(A_1))(\mathrm{tr}\exp(A_2)) \cdots \mathrm{tr}\exp(A_r)).$$

Theorem. Let A_1, A_2, B_1, B_2 be real $n \times n$ matrices and let I be the $n \times n$ identity matrix. Then

$$\mathrm{tr}\exp(A_1 \otimes I \otimes B_1 \otimes I + I \otimes A_2 \otimes I \otimes B_2) \equiv (\mathrm{tr}\exp(A_1 \otimes B_1))(\mathrm{tr}\exp(A_2 \otimes B_2)).$$

Proof. There is an $n^2 \times n^2$ permutation matrix P such that, for any $n \times n$ matrices M and N,

$$P(M \otimes N)P^T = N \otimes M.$$

Define an $n^4 \times n^4$ permutation matrix Q by

$$Q := I \otimes P \otimes I.$$

It follows that

$$Q(A_1 \otimes B_1 \otimes I \otimes I + I \otimes I \otimes A_2 \otimes B_2)Q^T \equiv A_1 \otimes I \otimes B_1 \otimes I + I \otimes A_2 \otimes I \otimes B_2.$$

Since $Q^T Q = I$ there is a similarity between the matrices

$$A_1 \otimes B_1 \otimes I \otimes I + I \otimes I \otimes A_2 \otimes B_2$$

and

$$A_1 \otimes I \otimes B_1 \otimes I + I \otimes A_2 \otimes I \otimes B_2.$$

So the two matrices have the same eigenvalues. Consequently, the same is true of the exponential function evaluated at these matrices and, in particular, they have the same trace. ♠

An extension of the formula given above is

Theorem. Let $A_1, A_2, \ldots, A_r, B_1, B_2, \ldots, B_r$ be real $n \times n$ matrices and let I be the $n \times n$ identity matrix. Then

$$\operatorname{tr}\exp(A_1 \otimes I \otimes \cdots \otimes I \otimes B_1 \otimes I \otimes \cdots \otimes I + I \otimes A_2 \otimes \cdots \otimes I \otimes I \otimes B_2 \otimes \cdots \otimes I + \cdots$$

$$+ I \otimes I \otimes \cdots \otimes A_r \otimes I \otimes I \otimes \cdots \otimes B_r)$$

$$= (\operatorname{tr}\exp(A_1 \otimes B_1))(\operatorname{tr}\exp(A_2 \otimes B_2)) \ldots (\operatorname{tr}\exp(A_r \otimes B_r)).$$

The proof of this theorem is completely parallel to that of theorem given above (Steeb [34]).

Let us now consider determinants.

Theorem. Let A be an $m \times m$ matrix and B be an $n \times n$ matrix. Then

$$\det(A \otimes B) \equiv (\det A)^n (\det B)^m.$$

Proof. Let I_m be the $m \times m$ unit matrix. Then

$$\det(I_m \otimes B) \equiv (\det B)^m$$

and

$$\det(A \otimes I_n) \equiv \det(P(I_n \otimes A)P^T) \equiv (\det P)^2 (\det A)^n \equiv (\det A)^n.$$

Now using

$$A \otimes B \equiv (A \otimes I_n)(I_m \otimes B)$$

we obtain

$$\det(A \otimes B) \equiv \det[(A \otimes I_n)(I_m \otimes B)] \equiv \det(A \otimes I_n)\det(I_m \otimes B) \equiv (\det A)^n (\det B)^m. \spadesuit$$

Remark. Another proof of this theorem with the help of the eigenvalues of $A \otimes B$ will be given in section 2.6.

Exercises. (1) Let P be an $n \times n$ permutation matrix. Show that

$$\det(P \otimes P) = 1.$$

(2) Let A be an $m \times m$ matrix and B be an $n \times n$ matrix. Show that

$$\det \exp(A \otimes B) \equiv \exp((\text{tr}A)(\text{tr}B)).$$

(3) Let A be an $n \times n$ idempotent matrix. Find $\det(A \otimes A)$.

(4) Let A be an $n \times n$ nilpotent matrix. Find $\det(A \otimes A)$.

(5) Let A be an $n \times n$ matrix over $\mathbf{R}$. Assume that A is skew symmetric. Let B be an $(n+1) \times (n+1)$ skew symmetric matrix over $\mathbf{R}$. Show that

$$\det(A \otimes B) = 0.$$

(6) Let U be a unitary matrix. Show that

$$\det(U \otimes U \otimes U \otimes U) = 1.$$

(7) Let A be an $n \times n$ matrix with $\det A = 1$. Show that

$$\det(A^{-1} \otimes A) = 1.$$

(8) Let A, B be $n \times n$ matrices. Assume that $[A, B]_+ = 0$, where $[\,,\,]_+$ denotes the anticommutator. Find $\det(AB)$.

(9) Let A, B be $n \times n$ matrices. Assume that

$$A \otimes B + B \otimes A = 0.$$

Find $\det(A \otimes B)$.

(10) Let A, B be $n \times n$ matrices. Show that

$$\text{tr}(A \otimes B - B \otimes A) = 0.$$

2.6 Eigenvalue Problem

Let A be an $m \times m$ matrix and let B be an $n \times n$ matrix. We now investigate the eigenvalue problem for the matrices $A \otimes B$ and $A \otimes I_n + I_m \otimes B$.

Theorem. Let A be an $m \times m$ matrix with eigenvalues $\lambda_1, \lambda_2, \cdots, \lambda_m$ and the corresponding eigenvectors $\mathbf{u}_1, \mathbf{u}_2, \cdots, \mathbf{u}_m$. Let B be an $n \times n$ matrix with eigenvalues $\mu_1, \mu_2, \cdots, \mu_n$ and the corresponding eigenvectors $\mathbf{v}_1, \mathbf{v}_2, \cdots, \mathbf{v}_n$. Then the matrix $A \otimes B$ has the eigenvalues $\lambda_j \mu_k$ with the corresponding eigenvectors $\mathbf{u}_j \otimes \mathbf{v}_k$, where $1 \leq j \leq m$ and $1 \leq k \leq n$.

Proof. From the eigenvalue equations

$$A\mathbf{u}_j = \lambda_j \mathbf{u}_j, \qquad 1 \leq j \leq m$$

$$B\mathbf{v}_k = \mu_k \mathbf{v}_k, \qquad 1 \leq k \leq n$$

we obtain

$$(A\mathbf{u}_j) \otimes (B\mathbf{v}_k) = \lambda_j \mu_k (\mathbf{u}_j \otimes \mathbf{v}_k).$$

Since

$$(A\mathbf{u}_j) \otimes (B\mathbf{v}_k) \equiv (A \otimes B)(\mathbf{u}_j \otimes \mathbf{v}_k)$$

we arrive at

$$(A \otimes B)(\mathbf{u}_j \otimes \mathbf{v}_k) = \lambda_j \mu_k (\mathbf{u}_j \otimes \mathbf{v}_k).$$

This equation is an eigenvalue equation. Consequently, $\mathbf{u}_j \otimes \mathbf{v}_k$ is an eigenvector of $A \otimes B$ with eigenvalue $\lambda_j \mu_k$. ♠

Example. Let

$$A = \begin{pmatrix} 0 & -i \\ i & 0 \end{pmatrix}, \qquad B = \begin{pmatrix} 1 & 0 \\ 0 & -1 \end{pmatrix}.$$

Since the eigenvalues of A and B are given by $\{1, \ -1\}$, we find that the eigenvalues of $A \otimes B$ are given by $\{1, \ 1, \ -1, \ -1\}$. ♣

Let A be an $m \times m$ matrix and let B be an $n \times n$ matrix. Then

$$\det(A \otimes B) = (\det A)^n (\det B)^m.$$

The proof can be given using the theorem described above. First we note that $\det(A) = \lambda_1 \lambda_2 \cdots \lambda_m$ and $\det(B) = \mu_1 \mu_2 \cdots \mu_n$. We find

$$\det(A \otimes B) = \prod_{j=1}^{m} \prod_{k=1}^{n} (\lambda_j \mu_k) = \left(\prod_{j=1}^{m} \lambda_j \right)^n \left(\prod_{k=1}^{n} \mu_k \right)^m = (\det A)^n (\det B)^m.$$

Theorem. Let A be an $m \times m$ matrix with eigenvalues $\lambda_1, \lambda_2, \cdots, \lambda_m$ and the corresponding eigenvectors $\mathbf{u}_1, \mathbf{u}_2, \cdots, \mathbf{u}_m$. Let B be an $n \times n$ matrix with eigenvalues $\mu_1, \mu_2, \cdots, \mu_n$ and the corresponding eigenvectors $\mathbf{v}_1, \mathbf{v}_2, \cdots, \mathbf{v}_n$. Then the eigenvalues and eigenvectors of

$$A \otimes I_n + I_m \otimes B$$

are given by $\lambda_j + \mu_k$ and $\mathbf{u}_j \otimes \mathbf{v}_k$, respectively, where $j = 1, 2, \cdots, m$ and $k = 1, 2, \cdots, n$, and I_m is the $m \times m$ unit matrix and I_n is the $n \times n$ unit matrix.

Proof. We have

$$\begin{aligned}
(A \otimes I_n + I_m \otimes B)(\mathbf{u}_j \otimes \mathbf{v}_k) &= (A \otimes I_n)(\mathbf{u}_j \otimes \mathbf{v}_k) + (I_m \otimes B)(\mathbf{u}_j \otimes \mathbf{v}_k) \\
&= (A\mathbf{u}_j) \otimes (I_n \mathbf{v}_k) + (I_m \mathbf{u}_j) \otimes (B\mathbf{v}_k) \\
&= (\lambda_j \mathbf{u}_j) \otimes \mathbf{v}_k + \mathbf{u}_j \otimes (\mu_k \mathbf{v}_k) \\
&= (\lambda_j + \mu_k)(\mathbf{u}_j \otimes \mathbf{v}_k). \quad \spadesuit
\end{aligned}$$

Example. Let

$$A = \begin{pmatrix} 0 & -i \\ i & 0 \end{pmatrix}, \qquad B = \begin{pmatrix} 1 & 0 \\ 0 & -1 \end{pmatrix}.$$

Since the eigenvalues of A and B are given by $\{1, -1\}$, we find that the eigenvalues of $A \otimes I_2 + I_2 \otimes B$ are given by $\{2, 0, 0, -2\}$. $\clubsuit$

The expression $A \otimes I_n + I_m \otimes B$ is sometimes called the *Kronecker sum*. The Kronecker sum plays a role when we consider the equation

$$AX + XB = C$$

where A is an $m \times m$ matrix, B is an $n \times n$ matrix, X is an $m \times n$ matrix and C is an $m \times n$ matrix. This equation can be written in the form

$$(I_n \otimes A + B^T \otimes I_m)\text{vec}X = \text{vec}C$$

where we used $\text{vec}(AYB) = (B^T \otimes A)\text{vec}Y$.

Exercises. (1) Let A and B be $n \times n$ matrices with eigenvalues $\lambda_1, \lambda_2, \ldots, \lambda_n$ and $\mu_1, \mu_2, \ldots, \mu_n$, respectively. Find the eigenvalues of $A \otimes B - B \otimes A$.

(2) Let A and B be $n \times n$ matrices with eigenvalues $\lambda_1, \lambda_2, \ldots, \lambda_n$ and $\mu_1, \mu_2, \ldots, \mu_n$, respectively. Find the eigenvalues of $A \otimes B + B \otimes A$.

(3) Let A, B, C be $n \times n$ matrices. $\lambda_1, \lambda_2, \ldots, \lambda_n$, $\mu_1, \mu_2, \ldots, \mu_n$ and $\nu_1, \nu_2, \ldots, \nu_n$, respectively. Find the eigenvalues of $A \otimes B \otimes C$.

(4) Let A be an $m \times m$ matrix with eigenvalue λ and eigenvector $\mathbf{u}$. Let B be an $n \times n$ matrix with eigenvalue μ and eigenvector $\mathbf{v}$. Show that

$$\exp(A \otimes B)(\mathbf{u} \otimes \mathbf{v}) \equiv \exp(\lambda\mu)(\mathbf{u} \otimes \mathbf{v}).$$

(5) Let A, B, C be $n \times n$ matrices. Let I be the $n \times n$ unit matrix. Find the eigenvalues of

$$A \otimes I \otimes I + I \otimes B \otimes I + I \otimes I \otimes C.$$

(6) Let A be an $n \times n$ matrix. Let λ_1, λ_2, ..., λ_n be the eigenvalues of A. Assume that A^{-1} exists. Show that

$$\mathrm{tr}(A^{-1}) = \sum_{j=1}^{n} \lambda_j^{-1}.$$

Find the eigenvalues of $A^{-1} \otimes A$.

(7) Let A be an $n \times n$ matrix with eigenvalues λ_1, λ_2, ..., λ_n. Let B be an $n \times n$ matrix with eigenvalues μ_1, μ_2, ..., μ_n. Let I be the $n \times n$ unit matrix and $\epsilon \in \mathbf{R}$. Show that

$$I \otimes I + \epsilon(A \otimes I + I \otimes B) + \epsilon^2(A \otimes B)$$

has the eigenvalues

$$(1 + \epsilon\lambda_j)(1 + \epsilon\mu_k).$$

(8) Let

$$A := (u_1, u_2) \otimes \begin{pmatrix} u_1 \\ u_2 \end{pmatrix}.$$

Show that at least one eigenvalue of A is equal to zero. Hint: Calculate $\det A$ and use the fact that $\det A = \lambda_1\lambda_2$.

(9) Let

$$A := (u_1, u_2, u_3) \otimes \begin{pmatrix} u_1 \\ u_2 \\ u_3 \end{pmatrix}.$$

Show that at least one eigenvalue of A is equal to zero. Hint: Calculate $\det A$ and use the fact that $\det A = \lambda_1 \lambda_2 \lambda_3$. Generalize to the n-dimensional case.

(10) Let $f(x, y)$ designate the polynomial

$$f(x, y) = \sum_{j=0}^{p} \sum_{k=0}^{p} a_{jk} x^j y^k.$$

Let A be an $m \times m$ matrix and B be an $n \times n$ matrix. Let $f(A, B)$ designate the $mn \times mn$ matrix

$$f(A, B) := \sum_{j=0}^{p} \sum_{k=0}^{p} a_{jk} A^j \otimes B^k.$$

Show that the eigenvalues of $f(A, B)$ are given by

$$f(\lambda_r, \mu_s), \qquad r = 1, 2, \ldots, m, \qquad s = 1, 2, \ldots, n$$

where λ_r and μ_s are the eigenvalues of A and B, respectively.

(11) Let A be an $n \times n$ matrix with eigenvalues $\lambda_1, \lambda_2, \ldots, \lambda_n$. Let I be the $n \times n$ unit matrix. Find the eigenvalues of $I \otimes A$, $A \otimes I$, $A \otimes I \otimes I$, $I \otimes A \otimes I$ and $I \otimes I \otimes A$. Extend to n-factors.

(12) Show that the eigenvalues of the $n \times n$ matrix

$$A = \begin{pmatrix} a & b & 0 & \ldots & 0 \\ b & a & b & \ldots & 0 \\ 0 & b & a & \ldots & 0 \\ \vdots & & \ddots & & b \\ 0 & \ldots & & b & a \end{pmatrix}$$

are given by

$$\lambda_j = a + 2b \cos\left(\frac{\pi j}{n+1}\right)$$

where $j = 1, 2, \ldots, n$. Find the eigenvalues of $A \otimes A$, $A \otimes I$ and $I \otimes A$.

2.7 Projection Matrices

We introduced projection matrices in section 1.4. Here we study the Kronecker product of projection matrices.

Theorem. Let Π be a $m \times m$ pojection matrix, Π_1 be a $n \times n$ projection matrix, Π_2 be a $p \times p$ projection matrix. Let I is the $r \times r$ unit matrix. Then $\Pi_1 \otimes \Pi_2$ and $\Pi \otimes I$ are projection matrices.

Proof. First we notice that

$$(\Pi_1 \otimes \Pi_2)^* = \Pi_1^* \otimes \Pi_2^* = \Pi_1 \otimes \Pi_2.$$

Since

$$(\Pi_1 \otimes \Pi_2)(\Pi_1 \otimes \Pi_2) = (\Pi_1^2) \otimes (\Pi_2^2) = \Pi_1 \otimes \Pi_2$$

we find that $\Pi_1 \otimes \Pi_2$ is a projection matrix. Owing to

$$(\Pi \otimes I)^* = (\Pi^* \otimes I^*) = \Pi \otimes I$$

and

$$(\Pi \otimes I)(\Pi \otimes I) = (\Pi^2 \otimes I^2) = \Pi \otimes I$$

we find that $\Pi \otimes I$ is a projection matrix. ♠

Theorem. Let Π_1 be an $m \times m$ projection matrix, Π_2 be an $n \times n$ projection matrix. Then $I_m \otimes I_n - \Pi_1 \otimes \Pi_2$ is a projection matrix.

Proof. We have

$$(I_m \otimes I_n - \Pi_1 \otimes \Pi_2)^* = (I_m^* \otimes I_n^* - \Pi_1^* \otimes \Pi_2^*) = (I_m \otimes I_n - \Pi_1 \otimes \Pi_2).$$

Furthermore

$$(I_m \otimes I_n - \Pi_1 \otimes \Pi_2)^2 = (I_m^2 \otimes I_n^2 - \Pi_1 \otimes \Pi_2 - \Pi_1 \otimes \Pi_2 + \Pi_1^2 \otimes \Pi_2^2).$$

Since $\Pi_1^2 = \Pi_1$, $\Pi_2^2 = \Pi_2$, $I_m^2 = I_m$ we proved the theorem. ♠

Exercises. (1) Let

$$\Pi_1 = \frac{1}{2}\begin{pmatrix} 1 & 1 \\ 1 & 1 \end{pmatrix}, \qquad \Pi_2 = \begin{pmatrix} 0 & 0 \\ 0 & 1 \end{pmatrix}.$$

Find $\Pi_1 \otimes \Pi_2$.

(2) Show that the matrices

$$\Pi_1 = \frac{1}{2}\begin{pmatrix} 1 & 1 \\ 1 & 1 \end{pmatrix}, \qquad \Pi_2 = \frac{1}{2}\begin{pmatrix} 1 & -1 \\ -1 & 1 \end{pmatrix}$$

are projection matrices and that $\Pi_1\Pi_2 = 0$. Show that

$$(\Pi_1 \otimes \Pi_2)(\Pi_2 \otimes \Pi_1) = 0\,.$$

(3) Let $\Pi_1, \cdots, \Pi_r$ be $n \times n$ projection matrices. Show that

$$\Pi_1 \otimes \Pi_2 \otimes \cdots \otimes \Pi_r$$

is a projection matrix.

(4) Let Π_1 and Π_2 be projection matrices. Find $\det(\Pi_1 \otimes \Pi_2)$.

(5) Let Π_1 and Π_2 be projection matrices. Find the eigenvalues of $\Pi_1 \otimes \Pi_2$.

(6) Let $\mathbf{u} \in \mathbf{C}^n$ with $(\mathbf{u}, \mathbf{u}) = 1$. Show that

$$(\mathbf{u}^* \otimes \mathbf{u})^2 = \mathbf{u}^* \otimes \mathbf{u}.$$

(7) Let $\mathbf{u}, \mathbf{v} \in \mathbf{C}^n$ and $(\mathbf{u}, \mathbf{v}) = 0$. Show that

$$(\mathbf{u}^* \otimes \mathbf{v})^2 = 0.$$

(8) Let Π be projection matrix and P be a permutation matrix. What is the condition on P such that $\Pi \otimes P$ is a projection matrix.

2.8 Spectral Representation

Here we discuss whether a Hermitian matrix can be reconstructed from its eigenvalues and eigenfunctions using the Kronecker product. This reconstruction is called *spectral representation* of a Hermitian matrix.

Before we describe the spectral representation let us consider two examples.

Example. Let

$$A = \begin{pmatrix} 0 & -i \\ i & 0 \end{pmatrix}.$$

The eigenvalues are

$$\lambda_1 = 1, \qquad \lambda_2 = -1$$

with the corresponding normalized eigenvectors

$$\mathbf{u}_1 = \frac{1}{\sqrt{2}} \begin{pmatrix} 1 \\ i \end{pmatrix}, \qquad \mathbf{u}_2 = \frac{1}{\sqrt{2}} \begin{pmatrix} 1 \\ -i \end{pmatrix}.$$

Then

$$\begin{aligned} A &\equiv \lambda_1 \mathbf{u}_1^* \otimes \mathbf{u}_1 + \lambda_2 \mathbf{u}_2^* \otimes \mathbf{u}_2 \\ &\equiv \frac{1}{2}(1,-i) \otimes \begin{pmatrix} 1 \\ i \end{pmatrix} - \frac{1}{2}(1,i) \otimes \begin{pmatrix} 1 \\ -i \end{pmatrix} \\ &\equiv \frac{1}{2}\begin{pmatrix} 1 & -i \\ i & 1 \end{pmatrix} - \frac{1}{2}\begin{pmatrix} 1 & i \\ -i & 1 \end{pmatrix} \equiv \begin{pmatrix} 0 & -i \\ i & 0 \end{pmatrix}. \quad \clubsuit \end{aligned}$$

Example. Let

$$A = \begin{pmatrix} 5 & -2 & -4 \\ -2 & 2 & 2 \\ -4 & 2 & 5 \end{pmatrix}.$$

Then the eigenvalues are

$$\lambda_1 = 1, \qquad \lambda_2 = 1, \qquad \lambda_3 = 10.$$

This means the eigenvalue $\lambda = 1$ is twofold. The eigenvectors are

$$\mathbf{u}_1 = \begin{pmatrix} -1 \\ -2 \\ 0 \end{pmatrix}, \qquad \mathbf{u}_2 = \begin{pmatrix} -1 \\ 0 \\ -1 \end{pmatrix}, \qquad \mathbf{u}_3 = \begin{pmatrix} 2 \\ -1 \\ -2 \end{pmatrix}.$$

We find that

$$(\mathbf{u}_1, \mathbf{u}_3) = 0, \qquad (\mathbf{u}_2, \mathbf{u}_3) = 0, \qquad (\mathbf{u}_1, \mathbf{u}_2) = 1.$$

However, the two vectors $\mathbf{u}_1$ and $\mathbf{u}_2$ are linearly independent. Now we use the Gram-Schmidt algorithm to find orthogonal eigenvectors.

The *Gram-Schmidt algorithm* is as follows: Let $\mathbf{v}_1, \mathbf{v}_2, \ldots, \mathbf{v}_n$ be a basis in $\mathbf{C}^n$. We define

$$\begin{aligned}
\mathbf{w}_1 &:= \mathbf{v}_1 \\
\mathbf{w}_2 &:= \mathbf{v}_2 - \frac{(\mathbf{w}_1, \mathbf{v}_2)}{(\mathbf{w}_1, \mathbf{w}_1)}\mathbf{w}_1 \\
\mathbf{w}_3 &:= \mathbf{v}_3 - \frac{(\mathbf{w}_2, \mathbf{v}_3)}{(\mathbf{w}_2, \mathbf{w}_2)}\mathbf{w}_2 - \frac{(\mathbf{w}_1, \mathbf{v}_3)}{(\mathbf{w}_1, \mathbf{w}_1)}\mathbf{w}_1 \\
\ldots & \ldots \ldots \\
\mathbf{w}_n &:= \mathbf{v}_n - \frac{(\mathbf{w}_{n-1}, \mathbf{v}_n)}{(\mathbf{w}_{n-1}, \mathbf{w}_{n-1})}\mathbf{w}_{n-1} - \ldots - \frac{(\mathbf{w}_1, \mathbf{v}_n)}{(\mathbf{w}_1, \mathbf{w}_1)}\mathbf{w}_1.
\end{aligned}$$

Then the vectors $\mathbf{w}_1, \mathbf{w}_2, \ldots, \mathbf{w}_n$ form an orthonormal basis in $\mathbf{C}^n$. In our example we choose

$$\mathbf{u}_1' = \mathbf{u}_1, \qquad \mathbf{u}_2' = \mathbf{u}_2 + \alpha\mathbf{u}_1$$

such that

$$\alpha := -\frac{(\mathbf{u}_1, \mathbf{u}_2)}{(\mathbf{u}_1, \mathbf{u}_1)} = -\frac{1}{5}.$$

Then

$$\mathbf{u}_2' = \frac{1}{5}\begin{pmatrix} -4 \\ 2 \\ -5 \end{pmatrix}.$$

The normalized eigenvectors are

$$\mathbf{u}_1 = \frac{1}{\sqrt{5}}\begin{pmatrix} -1 \\ -2 \\ 0 \end{pmatrix}, \qquad \mathbf{u}_2' = \frac{1}{3\sqrt{5}}\begin{pmatrix} -4 \\ 2 \\ -5 \end{pmatrix}, \qquad \mathbf{u}_3 = \frac{1}{3}\begin{pmatrix} 2 \\ -1 \\ -2 \end{pmatrix}.$$

Consequently

$$A = \lambda_1 \mathbf{u}_1^T \otimes \mathbf{u}_1 + \lambda_2 \mathbf{u}_2'^T \otimes \mathbf{u}_2' + \lambda_3 \mathbf{u}_3^T \otimes \mathbf{u}_3.$$

From the normalized eigenvectors we obtain the orthogonal matrix

$$T = \begin{pmatrix} \frac{-1}{\sqrt{5}} & \frac{-4}{3\sqrt{5}} & \frac{2}{3} \\ \frac{-2}{\sqrt{5}} & \frac{2}{3\sqrt{5}} & -\frac{1}{3} \\ 0 & \frac{-5}{3\sqrt{5}} & -\frac{2}{3} \end{pmatrix}.$$

Therefore

$$T^T A T = \begin{pmatrix} 1 & 0 & 0 \\ 0 & 1 & 0 \\ 0 & 0 & 10 \end{pmatrix}$$

where $T^T = T^{-1}$.

We now discuss the spectral representation of a Hermitian matrix.

Theorem. Let A be a $n \times n$ Hermitian matrix. Assume that the eigenvalues $\lambda_1, \cdots, \lambda_n$ are all distinct. Then the normalized eigenvectors $\mathbf{u}_1, \cdots, \mathbf{u}_n$ are orthonormal and form a basis in the vector space $\mathbf{C}^n$. We have

$$A = \sum_{j=1}^{n} \lambda_j \mathbf{u}_j^* \otimes \mathbf{u}_j.$$

Proof. Let $\mathbf{v}$ be an arbitary vector in $\mathbf{C}^n$. Then $\mathbf{v}$ can be expanded as

$$\mathbf{v} = \sum_{j=1}^{n} (\mathbf{v}, \mathbf{u}_j)\mathbf{u}_j \equiv \sum_{j=1}^{n} (\mathbf{v}^T \bar{\mathbf{u}}_j)\mathbf{u}_j$$

where $(\mathbf{v}, \mathbf{u}_j)$ denotes the scalar product in $\mathbf{C}^n$. Applying A to $\mathbf{v}$ gives

$$A\mathbf{v} = \sum_{j=1}^{n} (\mathbf{v}, \mathbf{u}_j) A\mathbf{u}_j = \sum_{j=1}^{n} (\mathbf{v}, \mathbf{u}_j)\lambda_j \mathbf{u}_j = \sum_{j=1}^{n} \lambda_j (\mathbf{v}, \mathbf{u}_j)\mathbf{u}_j.$$

Thus we have to prove that

$$\left(\sum_{j=1}^{n} \lambda_j \mathbf{u}_j^* \otimes \mathbf{u}_j \right) \mathbf{v} \equiv \sum_{j=1}^{n} \lambda_j (\mathbf{v}, \mathbf{u}_j)\mathbf{u}_j$$

or

$$(\mathbf{u}_j^* \otimes \mathbf{u}_j)\mathbf{v} \equiv (\mathbf{v}, \mathbf{u}_j)\mathbf{u}_j$$

since $\lambda_j \neq \lambda_k$ for $j \neq k$. We set

$$\mathbf{v}^T = (v_1, v_2, \ldots, v_n), \qquad \mathbf{u}_j^T = (u_{j1}, u_{j2}, \ldots, u_{jn}).$$

Since

$$\begin{aligned}
(\mathbf{u}_j^* \otimes \mathbf{u}_j)\mathbf{v} &= \begin{pmatrix} \bar{u}_{j1}u_{j1} & \bar{u}_{j2}u_{j1} & \cdots & \bar{u}_{jn}u_{j1} \\ \bar{u}_{j1}u_{j2} & \bar{u}_{j2}u_{j2} & & \bar{u}_{jn}u_{j2} \\ \vdots & \vdots & & \\ \bar{u}_{j1}u_{jn} & \bar{u}_{j2}u_{jn} & & \bar{u}_{jn}u_{jn} \end{pmatrix} \begin{pmatrix} v_1 \\ v_2 \\ \vdots \\ v_n \end{pmatrix} \\
&\equiv \begin{pmatrix} \bar{u}_{j1}u_{j1}v_1 & + \bar{u}_{j2}u_{j1}v_2 & + \cdots & + \bar{u}_{jn}u_{j1}v_n \\ \bar{u}_{j1}u_{j2}v_1 & + \bar{u}_{j2}u_{j2}v_2 & + \cdots & + \bar{u}_{jn}u_{j2}v_n \\ \vdots & & & \vdots \\ \bar{u}_{j1}u_{jn}v_1 & + \bar{u}_{j2}u_{jn}v_2 & + \cdots & + \bar{u}_{jn}u_{jn}v_n \end{pmatrix}
\end{aligned}$$

and

$$
\begin{aligned}
(\mathbf{v}, \mathbf{u}_j)\mathbf{u}_j &= \sum_{k=1}^{n} (v_k \bar{u}_{jk})\mathbf{u}_j \\
&= \begin{pmatrix} v_1\bar{u}_{j1}u_{j1} & + v_2\bar{u}_{j2}u_{j1} & + \cdots & + v_n\bar{u}_{jn}u_{j1} \\ v_1\bar{u}_{j1}u_{j2} & + v_2\bar{u}_{j2}u_{j2} & + \cdots & + v_n\bar{u}_{jn}u_{j2} \\ \vdots & & & \vdots \\ v_1\bar{u}_{j1}u_{jn} & + v_2\bar{u}_{j2}u_{jn} & + \cdots & + v_n\bar{u}_{jn}u_{jn} \end{pmatrix}
\end{aligned}
$$

we have proved the spectral representation. ♠

The above theorem can also be applied when A has multiple eigenvalues, but with corresponding orthogonal eigenvectors.

If on the other hand the Hermitian matrix A has multiple eigenvalues with corresponding non-orthogonal eigenvectors, we proceed as follows. Let λ be an eigenvalue of multiplicity m. Then the eigenvalues with their corresponding eigenvectors can be ordered as

$$
\lambda,\ \lambda, \ldots, \lambda,\ \lambda_{m+1}, \ldots, \lambda_n, \qquad \mathbf{u}_1,\ \mathbf{u}_2, \ldots, \mathbf{u}_m,\ \mathbf{u}_{m+1}, \ldots, \mathbf{u}_n.
$$

The vectors $\mathbf{u}_{m+1}, \ldots, \mathbf{u}_n$ are orthogonal to each other and to the rest. What is left is to find a new set of orthogonal vectors $\mathbf{u}'_1, \mathbf{u}'_2, \ldots, \mathbf{u}'_m$ each being orthogonal to $\mathbf{u}_{m+1}, \ldots, \mathbf{u}_n$ together with each being an eigenvector of A. The procedure we use is the Gram-Schmidt algorithm described above.

Let $\mathbf{u}'_1 = \mathbf{u}_1$ and $\mathbf{u}'_2 = \mathbf{u}_2 + \alpha\mathbf{u}_1$. Then $\mathbf{u}'_2$ is an eigenvector of A, for it is a combination of eigenvectors corresponding to the same eigenvalue λ. Also $\mathbf{u}'_2$ is orthogonal to $\mathbf{u}_{m+1}, \ldots, \mathbf{u}_n$ since the latter are orthogonal to $\mathbf{u}_1$ and $\mathbf{u}_2$. What remains is to make $\mathbf{u}'_2$ orthogonal to $\mathbf{u}'_1$ i.e. to $\mathbf{u}_1$. We obtain

$$
\alpha = -\frac{(\mathbf{u}_1, \mathbf{u}_2)}{(\mathbf{u}_1, \mathbf{u}_1)}.
$$

Next we set

$$
\mathbf{u}'_3 = \mathbf{u}_3 + \alpha\mathbf{u}_1 + \beta\mathbf{u}_2
$$

where α and β have to be determined. Using the same reasoning, we obtain the linear equation for α and β

$$
\begin{pmatrix} (\mathbf{u}_1, \mathbf{u}_1) & (\mathbf{u}_1, \mathbf{u}_2) \\ (\mathbf{u}_2, \mathbf{u}_1) & (\mathbf{u}_2, \mathbf{u}_2) \end{pmatrix} \begin{pmatrix} \alpha \\ \beta \end{pmatrix} = -\begin{pmatrix} (\mathbf{u}_1, \mathbf{u}_3) \\ (\mathbf{u}_2, \mathbf{u}_3) \end{pmatrix}.
$$

The approach can be repeated until we obtain $\mathbf{u}'_m$. The Gramian matrix of the above equations is nonsingular, since the eigenvectors of a Hermitian matrix are linearly independent.

Exercises. (1) Consider the symmetric matrix

$$A = \begin{pmatrix} 1 & 1 & 1 & 1 \\ 1 & 1 & 1 & 1 \\ 1 & 1 & 1 & 1 \\ 1 & 1 & 1 & 1 \end{pmatrix}.$$

Find all eigenvalues and normalized eigenvectors. Reconstruct the matrix from the eigenvalues and normalized eigenvectors.

(2) Let A be a 4×4 symmetric matrix over $\mathbf{R}$ with eigenvalues $\lambda_1 = 0$, $\lambda_2 = 1$, $\lambda_3 = 2$, $\lambda_4 = 3$ and the corresponding normalized eigenvectors

$$\mathbf{u}_1 = \frac{1}{\sqrt{2}}(1, 0, 0, 1)^T$$

$$\mathbf{u}_2 = \frac{1}{\sqrt{2}}(1, 0, 0, -1)^T$$

$$\mathbf{u}_3 = \frac{1}{\sqrt{2}}(0, 1, 1, 0)^T$$

$$\mathbf{u}_4 = \frac{1}{\sqrt{2}}(0, 1, -1, 0)^T.$$

Find the matrix A.

(3) Let A be the skew symmetric matrix

$$A = \begin{pmatrix} 0 & 1 \\ -1 & 0 \end{pmatrix}.$$

Find the eigenvalues λ_1 and λ_2 of A and the corresponding normalized eigenvectors $\mathbf{u}_1$ and $\mathbf{u}_2$. Show that A is given by

$$A = \lambda_1 \mathbf{u}_1^* \otimes \mathbf{u}_1 + \lambda_2 \mathbf{u}_2^* \otimes \mathbf{u}_2.$$

(4) Explain why the matrix

$$A = \begin{pmatrix} 0 & 1 & 1 \\ 0 & 0 & 1 \\ 0 & 0 & 0 \end{pmatrix}$$

cannot be reconstructed from the eigenvalues and eigenvectors.

2.9 Fourier and Hadamard Matrices

In section 1.5 we introduced Fourier and Hadamard matrices. Here we discuss Fourier matrices and its connnection with the Kronecker product. The Fourier matrices of orders 2^n may be expressed as Kronecker products. This factorization is a manifestation, essentially, of the idea known as the *Fast Fourier Transform.*

Let F'_{2^n} denote the Fourier matrices of order 2^n whose rows have been permuted according to the bit reversing permutation.

Definition. A sequence in natural order can be arranged in *bit-reversed order* as follows: for an integer expressed in binary notation, reverse the binary form and transform to decimal notation, which is then called bit-reveresed notation.

Since the sequence 0, 1 is the bit reversed order of 0, 1 and 0, 2, 1, 3 is the bit reversed order of 0, 1, 2, 3 we find that the matrices F'_2 and F'_4 are given by

$$F'_2 = \frac{1}{\sqrt{2}}\begin{pmatrix} 1 & 1 \\ 1 & -1 \end{pmatrix} = F_2$$

$$F'_4 = \frac{1}{\sqrt{4}}\begin{pmatrix} 1 & 1 & 1 & 1 \\ 1 & -1 & 1 & -1 \\ 1 & i & -1 & -i \\ 1 & -i & -1 & i \end{pmatrix}.$$

We find that F'_4 can be written as

$$F'_4 = (I_2 \otimes F'_2)D_4(F'_2 \otimes I_2)$$

where D_4 is the 4×4 diagonal matrix

$$D_4 := \mathrm{diag}(1,1,1,i).$$

Since

$$A \otimes B = P(B \otimes A)P^T$$

for some permutation matrix P that depends merely on the dimensions of A and B, we may write, for some 4×4 permutation matrix P_4

$$F'_4 = (I_2 \otimes F'_2)D_4P_4(I_2 \otimes F'_2)P_4$$

where $P_4^{-1} = P_4$. Similarly,

$$F'_{16} = (I_4 \otimes F'_4)D_{16}(F'_4 \otimes I_4)$$

where D_{16} is the 16×16 diagonal matrix

$$D_{16} := \mathrm{diag}(I, D^2, D, D^3)$$

with

$$D := \mathrm{diag}(1, w, w^2, w^3), \qquad w := \exp\left(\frac{-2\pi i}{16}\right).$$

For an appropriate permutation matrix $P_{16} = P_{16}^{-1} = P_{16}^T$

$$F'_{16} = (I_4 \otimes F'_4) D_{16} P_{16} (I_4 \otimes F'_4) P_{16}.$$

For D_{256} we set

$$D_{256} = \mathrm{diag}(I, D^8, D^4, \cdots, D^{15})$$

where the sequence $0, 8, 4, 12, 2, 10, 6, 14, 1, 9, 5, 13, 3, 11, 7, 15$ is the bit reversed order of $0, 1, 2, \cdots, 15$ and

$$D := \mathrm{diag}(1, w, \cdots, w^{15}), \qquad w := \exp(2\pi i/256).$$

Development of reduced multiplication fast Fourier transform from small N discrete Fourier transform can be accomplished using Kronecker product expansions. Let $F_L, \ldots,$ F_2, F_1 be small N discrete Fourier transform algorithms with naturally ordered indices. Their dimensions are $N_L \times N_L, \ldots, N_2 \times N_2, N_1 \times N_1$, respectively. Then we can form the Kronecker products to be the $N \times N$ matrix F, where $N = N_L \cdots N_2 N_1$ and

$$F := F_L \otimes \cdots \otimes F_2 \otimes F_1.$$

For more details of this construction we refer to Elliott and Rao [13].

The *Winograd-Fourier transform* algorithm minimizes the number of multiplications, but not the number of additions, required to evaluate the reduced multiplication fast Fourier transform. The Winograd-Fourier transform algorithm (Elliott and Rao [13]) results from a Kronecker product manipulation to group input additions so that all transform multiplications follows. One writes

$$F = (S_L C_L T_L) \otimes \cdots \otimes (S_2 C_2 T_2) \otimes (S_1 C_1 T_1)$$

as

$$F = (S_L \otimes \cdots \otimes S_2 \otimes S_1)(C_L \otimes \cdots \otimes C_2 \otimes C_1)(T_L \otimes \cdots \otimes T_2 \otimes T_1)$$

where $(S_L \otimes \cdots \otimes S_2 \otimes S_1)$ describes the output additions, $(C_L \otimes \cdots \otimes C_2 \otimes C_1)$ the multiplications and $(T_L \otimes \cdots \otimes T_2 \otimes T_1)$ the input additions.

Hadamard matrices can also be constructed using the Kronecker product.

Theorem. If A and B are Hadamard matrices of orders m and n respectively, then $A \otimes B$ is an Hadamard matrix of order mn.

Proof. We have

$$(A \otimes B)(A \otimes B)^T = (A \otimes B)(A^T \otimes B^T) = (AA^T) \otimes (BB^T).$$

Thus

$$(A \otimes B)(A \otimes B)^T = (mI_m) \otimes (nI_n) = mn(I_m \otimes I_n) = mnI_{mn}. \quad \spadesuit$$

Sometimes the term Hadamard matrix is limited to the matrices of order 2^n given specifically by the recursion

$$\begin{aligned} H_1 &= (1) \\ H_2 &= \begin{pmatrix} 1 & 1 \\ 1 & -1 \end{pmatrix} \\ H_{2^{n+1}} &= H_{2^n} \otimes H_2. \end{aligned}$$

For example, we find

$$H_4 = \begin{pmatrix} 1 & 1 & 1 & 1 \\ 1 & -1 & 1 & -1 \\ 1 & 1 & -1 & -1 \\ 1 & -1 & -1 & 1 \end{pmatrix}.$$

These matrices have the property

$$H_{2^n} = H_{2^n}^T$$

so that

$$H_{2^n}^2 = 2^n I.$$

Definition. The *Walsh-Hadamard transform* is defined as

$$\hat{Z} = HZ$$

where H is an Hadamard matrix, where $Z = (z_1, z_2, \ldots, z_n)^T$. Since H is invertible the inverse transformation can be found

$$Z = H^{-1}\hat{Z}.$$

Closely related to the Hadamard matrices are the *Haar matrices.* The Haar transform is based on the Haar functions, which are periodic, orthogonal, and complete in the Hilbert space $L_2[0,1]$. The first two Haar matrices are given by

$$Ha(1) = \begin{pmatrix} 1 & 1 \\ 1 & -1 \end{pmatrix}$$

$$Ha(2) = \begin{pmatrix} 1 & 1 & 1 & 1 \\ 1 & 1 & -1 & -1 \\ \sqrt{2}(1 & -1 & 0 & 0) \\ \sqrt{2}(0 & 0 & 1 & -1) \end{pmatrix}.$$

Higher order Haar matrices can be generated recursively using the Kronecker product as follows

$$Ha(k+1) = \begin{pmatrix} Ha(k) & \otimes & (1 & 1) \\ 2^{k/2} I_{2^k} & \otimes & (1 & -1) \end{pmatrix} \qquad k > 1.$$

For example we find

$$Ha(3) = \begin{pmatrix} 1 & 1 & 1 & 1 & 1 & 1 & 1 & 1 \\ 1 & 1 & 1 & 1 & -1 & -1 & -1 & -1 \\ \sqrt{2}(1 & 1 & -1 & -1 & 0 & 0 & 0 & 0) \\ \sqrt{2}(0 & 0 & 0 & 0 & 1 & 1 & -1 & -1) \\ 2 & -2 & 0 & 0 & 0 & 0 & 0 & 0 \\ 0 & 0 & 2 & -2 & 0 & 0 & 0 & 0 \\ 0 & 0 & 0 & 0 & 2 & -2 & 0 & 0 \\ 0 & 0 & 0 & 0 & 0 & 0 & 2 & -2 \end{pmatrix}.$$

The Haar transform is given by

$$\mathbf{X} = \frac{1}{N}[Ha(L)]\mathbf{x}, \qquad \mathbf{x} = [Ha(L)]^T \mathbf{X}$$

where the $N \times N$ Haar matrix $Ha(L)$ is orthogonal

$$Ha(L)Ha(L)^T = NI_N.$$

Haar matrices can be factored into sparse matrices, which lead to the fast algorithms. Based on these algorithms both the Haar transform and its inverse can be implemented in $2(N-1)$ additions or subtractions and N multiplications.

2.10 Direct Sum

Let A be an $m \times n$ matrix and B be a $p \times q$ matrix. Then the *direct sum* of A and B is defined as

$$A \oplus B := \begin{pmatrix} A & 0 \\ 0 & B \end{pmatrix}.$$

All zero matrices are of appropriate order.

Example.

$$\begin{pmatrix} 1 & 0 & 3 \end{pmatrix} \oplus \begin{pmatrix} 6 & 7 \\ 8 & 9 \end{pmatrix} = \begin{pmatrix} 1 & 0 & 3 & 0 & 0 \\ 0 & 0 & 0 & 6 & 7 \\ 0 & 0 & 0 & 8 & 9 \end{pmatrix}. \quad \clubsuit$$

The extension to more than two matrices is straightforward, i. e.

$$A \oplus B \oplus C = \begin{pmatrix} A & 0 & 0 \\ 0 & B & 0 \\ 0 & 0 & C \end{pmatrix}$$

and

$$\bigoplus_{j=1}^{k} A_j = \begin{pmatrix} A_1 & 0 & 0 & \cdots & 0 \\ 0 & A_2 & 0 & \cdots & 0 \\ 0 & 0 & \ddots & & \vdots \\ \vdots & & & \ddots & 0 \\ 0 & 0 & \cdots & 0 & A_k \end{pmatrix} = \mathrm{diag}\{A_j\} \qquad \text{for } j = 1, \ldots, k.$$

The definition and its extensions apply whether or not the submatrices are the same order. All zero (null) matrices are of appropriate order.

Transposing a direct sum gives the direct sum of the transposes. The rank of a direct sum is the sum of the ranks, as is evident from the definition of rank. It is obvious that $A \oplus (-A) \neq 0$ unless A is null. Also,

$$(A \oplus B) + (C \oplus D) = (A + C) \oplus (B + D)$$

only if the necessary conditions of conformability for addition are met. Similarly,

$$(A \oplus B)(C \oplus D) = (AC) \oplus (BD)$$

provided that the matrix products exist. Let A and B be nonsingular $n \times n$ and $m \times m$ matrices, respectively. Then

$$(A \oplus B)^{-1} = A^{-1} \oplus B^{-1}.$$

The direct sum $A \oplus B$ is square only if A is $p \times q$ and B is $q \times p$. Let A and B be square matrices. Then

$$\det(A \oplus B) = \det(A)\det(B).$$

In general, we have

$$\det\left(\bigoplus_{i=1}^{N} A_i\right) = \prod_{i=1}^{N} \det(A_i).$$

and

$$r\left(\bigoplus_{i=1}^{N} A_i\right) = \sum_{i=1}^{N} r(A_i)$$

where $r(.)$ denotes the rank. If A and B are orthogonal, then $A \oplus B$ is orthogonal.

Theorem. Let A, B, C be arbitrary matrices. Then

$$(A \oplus B) \otimes C = (A \otimes C) \oplus (B \otimes C).$$

The proof is straightforward and is left as an exercise to the reader.

The Kronecker product of some matrices can be written as the direct sum of matrices.

Example. An example is

$$\begin{pmatrix} 1 & 0 \\ 0 & 1 \end{pmatrix} \otimes \begin{pmatrix} 1 & 1 \\ 1 & 1 \end{pmatrix} \equiv \begin{pmatrix} 1 & 1 \\ 1 & 1 \end{pmatrix} \oplus \begin{pmatrix} 1 & 1 \\ 1 & 1 \end{pmatrix}. \quad \clubsuit$$

This plays an important role in representation theory of groups (Ludwig and Falter [23]).

Exercises. (1) Show that in general

$$A \otimes (B \oplus C) \neq (A \otimes B) \oplus (A \otimes C).$$

Give an example.

(2) Let A be an $m \times m$ matrix with eigenvalues $\lambda_1, \lambda_2, \ldots, \lambda_m$. Let B be an $n \times n$ matrix with eigenvalues $\mu_1, \mu_2, \ldots, \mu_n$. Find the eigenvalues of $A \oplus B$.

(3) Let A be an $m \times m$ matrix with eigenvalues $\lambda_1, \lambda_2, \ldots, \lambda_m$. Let B be an $n \times n$ matrix with eigenvalues $\mu_1, \mu_2, \ldots, \mu_n$. Let C be an $r \times r$ matrix with eigenvalues $\nu_1, \nu_2, \ldots, \nu_r$. Find the eigenvalues of $(A \oplus B) \otimes C$.

(4) Let A, B, C and D be 2×2 matrices. Find the conditions on A, B, C and D such that

$$A \otimes B = C \oplus D.$$

(5) Let A, B, C be 2×2 matrices. Let D be a 3×3 matrix and E be a 2×2 matrix. Find the conditions on A, B, C, D, E such that

$$A \oplus B \oplus C = D \otimes E.$$

(6) Let P be a permutation matrix. Is $P \oplus P$ be a permutation matrix ? Is $(P \oplus P) \otimes P$ a permutation matrix ?

(7) Let Π be a projection matrix. Is $\Pi \oplus \Pi$ a projection matrix ? Is $(\Pi \oplus \Pi) \otimes \Pi$ a projection matrix ?

(8) Let A, B be Hermitian matrices. Is $A \oplus B$ a Hermitian matrix ?

2.11 Vec Operator

In section 1.8 we have introduced the vec operator. Here we investigate the connection of the vec operator with the Kronecker product. A comprehensive discussion is given by Graham [14] and Barnett [3].

The following theorems give useful properties of the vec operator. Proofs of the first two depend on the elementary vector $\mathbf{e}_j$, the j-th column of a unit matrix, i. e.

$$\mathbf{e}_1 = \begin{pmatrix} 1 \\ 0 \\ \vdots \\ 0 \end{pmatrix}, \qquad \mathbf{e}_2 = \begin{pmatrix} 0 \\ 1 \\ \vdots \\ 0 \end{pmatrix}, \ldots, \mathbf{e}_n = \begin{pmatrix} 0 \\ 0 \\ \vdots \\ 1 \end{pmatrix}.$$

Theorem. Let A, B, C be matrices such that the matrix product ABC exists. Then

$$\text{vec}(ABC) = (C^T \otimes A)\text{vec}B.$$

Proof. The j-th column of ABC, and hence the j-th subvector of vec(ABC) is, for C having r rows,

$$ABC\mathbf{e}_j = AB\sum_i \mathbf{e}_i\mathbf{e}_i^T C\mathbf{e}_j = \sum_i c_{ij}AB\mathbf{e}_i = (c_{1j}A \; c_{2j}A \cdots c_{rj}A)\begin{pmatrix} B\mathbf{e}_1 \\ B\mathbf{e}_2 \\ \vdots \\ B\mathbf{e}_r \end{pmatrix}$$

which is the j-th subvector of $(C^T \otimes A)\text{vec}B$. ♠

Corollary. Let A, B be $n \times n$ matrices. Then

$$\text{vec}(AB) = (I_n \otimes A)\text{vec}B$$

$$\text{vec}(AB) = (B^T \otimes A)\text{vec}I$$

$$\text{vec}(AB) = \sum_{k=1}^{n} (B^T)_k \otimes A_k$$

where A_k denotes the k-th column of the matrix A.

The proof is straightforward and left as an exercise to the reader.

Theorem. Let A, B, C, D be $n \times n$ matrices. Then

$$\operatorname{tr}(AD^TBDC) = (\operatorname{vec}D)^T(CA \otimes B^T)\operatorname{vec}D.$$

Proof.

$$\begin{aligned}\operatorname{tr}(AD^TBDC) &= \operatorname{tr}(D^TBDCA)\\ &= (\operatorname{vec}D)^T\operatorname{vec}(BDCA)\\ &= (\operatorname{vec}D)^T(A^TC^T \otimes B)\operatorname{vec}D\end{aligned}$$

and, being a scalar, this equals its own transpose. ♠

Consider the following equation

$$AX + XB = C$$

where A is an $n \times n$ matrix, B an $m \times m$ matrix and X an $n \times m$ matrix. Assume that A, B and C are given. The task is to find the matrix X. To solve this problem we apply the vec operator to the left and right-hand side of the equation given above. Thus the equation can be written in the form

$$(I_m \otimes A + B^T \otimes I_n)\operatorname{vec}X = \operatorname{vec}C.$$

This can easily be seen by applying the vec operator to both sides of $AX + XB = C$. Thus the equation $AX + XB = C$ has a unique solution if and only if the matrix $(I_m \otimes A + B^T \otimes I_n)$ is nonsingular. It follows that all the eigenvalues must be nonzero. The eigenvalues are of the form $\lambda_i + \mu_j$, where λ_i are the eigenvalues of A and μ_j are the eigenvalues of B. Thus the equation has a unique solution if and only if

$$\lambda_i + \mu_j \neq 0, \qquad \text{for all } i,\ j.$$

Thus we showed that $AX + XB = C$ has a unique solution if and only if A and $-B$ have no common eigenvalue.

Example. Let

$$A = B = \begin{pmatrix} 0 & 1 \\ 1 & 0 \end{pmatrix}.$$

Then the eigenvalues of A and $-B$ are the same, namely ± 1. Thus the equation $AX + XB = C$ does not have a unique solution. ♣

Exercises. (1) Let L, X, N, Y be $n \times n$ matrices. Assume that

$$LX + XN = Y.$$

Show that

$$(I \otimes L + N^T \otimes I)\text{vec}X = \text{vec}Y.$$

(2) Let $\mathbf{u}, \mathbf{v} \in \mathbf{C}^n$. Show that

$$\text{vec}(\mathbf{u}\mathbf{v}^T) = \mathbf{v} \otimes \mathbf{u}.$$

(3) Show that

$$\text{vec}(A \otimes \mathbf{u}) = (\text{vec}A) \otimes \mathbf{u}.$$

(4) Let

$$A = \begin{pmatrix} 1 & -1 \\ 0 & 2 \end{pmatrix}, \qquad B = \begin{pmatrix} 2 & -1 \\ 1 & 1 \end{pmatrix}, \qquad C = \begin{pmatrix} 1 & 3 \\ 2 & 2 \end{pmatrix}.$$

Find the solution to $AX + XB = C$.

(5) Let A, B, C, and X be $n \times n$ matrices over $\mathbf{R}$. Solve the linear differential equation

$$\frac{dX}{dt} = AX + XB, \qquad X(t=0) = C$$

using the vec operator. As an example consider the matrices given by (4).

Hint: The solution is given by $X(t) = \exp(tA)C\exp(tB)$.

(6) Let A, B and X be $n \times n$ matrices over $\mathbf{R}$. Show that the differential equation

$$\frac{dX}{dt} = AXB, \qquad X(t=0) = X(0)$$

can be transformed into

$$\frac{d}{dt}\text{vec}(X) = (A \otimes B^T)\text{vec}(X).$$

Show that the general soluion is given by

$$\text{vec}(X(t)) = \exp(t(A \otimes B^T))\text{vec}(X(0)).$$

(7) Find the condition for the equation $AX - XA = \mu X$, $\mu \in \mathbf{C}$ to have a nontrivial solution.

2.12 Groups

In section 1.11 we introduced groups and some of their properties. Here we study the connection with the Kronecker product. In the following we consider groups which are given as $n \times n$ matrices with nonzero determinant and the matrix multiplication is the group multiplication. We start with an example.

Example. Let

$$e = \begin{pmatrix} 1 & 0 \\ 0 & 1 \end{pmatrix}, \qquad a = \begin{pmatrix} 0 & 1 \\ 1 & 0 \end{pmatrix}.$$

Then

$$ee = e, \qquad ea = a, \qquad ae = a, \qquad aa = e.$$

Obviously $\{ e, a \}$ forms a group under matrix multiplication. The group is abelian. Now we can ask whether

$$\{ e \otimes e, \; e \otimes a, \; a \otimes e, \; a \otimes a \}$$

form a group under matrix multiplication. We find

$$(e \otimes e)(e \otimes e) = e \otimes e, \qquad (e \otimes a)(e \otimes e) = e \otimes a$$

$$(e \otimes e)(e \otimes a) = e \otimes a, \qquad (e \otimes a)(e \otimes a) = e \otimes e$$

$$(a \otimes e)(e \otimes a) = a \otimes a, \qquad (e \otimes a)(a \otimes e) = a \otimes a$$

$$(a \otimes e)(a \otimes e) = e \otimes e, \qquad (a \otimes a)(a \otimes a) = e \otimes e$$

and

$$(e \otimes e)^{-1} = e \otimes e, \quad (e \otimes a)^{-1} = e \otimes a, \quad (a \otimes e)^{-1} = a \otimes e, \quad (a \otimes a)^{-1} = a \otimes a.$$

Thus, the set $\{ e \otimes e, e \otimes a, a \otimes e, a \otimes a \}$ forms a group under matrix multiplication. ♣

We consider now the general case. Let G be a group represented as $n \times n$ matrices and let G' be a group represented by $m \times m$ matrices. If $g_1, g_2 \in G$ and $g_1', g_2' \in G'$, then we have

$$(g_1 \otimes g_1')(g_2 \otimes g_2') = (g_1 g_2) \otimes (g_1' g_2').$$

We find that all pairs $g \otimes g'$ with $g \in G$ and $g' \in G'$ form a group. The identity element is

$$e \otimes e'$$

where e is the identity element of G (i.e. the $n \times n$ unit matrix) and e' is the identity element of G' (i.e the $m \times m$ unit matrix). The inverse of $g \otimes g'$ is given by

$$(g \otimes g')^{-1} = g^{-1} \otimes g'^{-1}.$$

Since

$$(g_1g_2)g_3 = g_1(g_2g_3)$$

and

$$(g'_1g'_2)g'_3 = g'_1(g'_2g'_3)$$

we have

$$\begin{aligned}
[(g_1 \otimes g'_1)(g_2 \otimes g'_2)]\,(g_3 \otimes g'_3) &= (g_1g_2 \otimes g'_1g'_2)(g_3 \otimes g'_3) \\
&= ((g_1g_2)g_3 \otimes (g'_1g'_2)g'_3) \\
&= (g_1(g_2g_3) \otimes g'_1(g'_2g'_3)) \\
&= (g_1 \otimes g'_1)\,[(g_2 \otimes g'_2)(g_3 \otimes g'_3)]\,.
\end{aligned}$$

The above consideration can be formulated in group theory (Miller [24]) in a more abstract way.

Let G and G' be two groups.

Definition. The *direct product* $G \times G'$ is the group consisting of all ordered pairs (g, g') with $g \in G$ and $g' \in G'$. The product of two group elements is given by

$$(g_1, g'_1)(g_2, g'_2) := (g_1g_2, g'_1g'_2).$$

Obviously, $G \times G'$ is a group with identity element (e, e') where e, e' are the identity elements of G, G', respectively. Indeed

$$(g, g')^{-1} = (g^{-1}, g'^{-1})$$

and the associative law is trivial to verify. The subgroup

$$G \times \{e'\} = \{(g, e') \,:\, g \in G\}$$

of $G \times G'$ is isomorphic to G with the isomorphism given by

$$(g, e') \leftrightarrow g\,.$$

Similarly the subgroup $\{e\} \times G'$ is isomorphic to G'. Since

$$(g, e')(e, g') = (e, g')(g, e') = (g, g')$$

it follows that (1) the elements of $G \times \{e'\}$ commute with the elements of $\{e\} \times G'$ and (2) every element of $G \times G'$ can be written uniquely as a product of an element in $g \times \{e'\}$ and an element in $\{e\} \times G'$.

Example. Let

$$e = \begin{pmatrix} 1 & 0 \\ 0 & 1 \end{pmatrix}, \qquad a = \begin{pmatrix} 0 & 1 \\ 1 & 0 \end{pmatrix}.$$

Then

$$ee = e, \qquad ea = a, \qquad ae = a, \qquad aa = e.$$

Obviously the set $\{ e, a \}$ forms a group under matrix multiplication. Now we can ask whether

$$\{ e \otimes e, \, a \otimes a \}$$

forms a group under matrix multiplication. We find

$$(e \otimes e)(e \otimes e) = e \otimes e, \qquad (e \otimes e)(a \otimes a) = a \otimes a$$

$$(a \otimes a)(e \otimes e) = a \otimes a, \qquad (a \otimes a)(a \otimes a) = e \otimes e$$

and

$$(e \otimes e)^{-1} = e \otimes e, \qquad (a \otimes a)^{-1} = a \otimes a\,.$$

The associative law is obviously satisfied. Thus we find that $e \otimes e, a \otimes a$ form a group under matrix multiplication. ♣

Obviously, the group $\{ e\otimes e \, , \, a\otimes a \}$ is a subgroup of the group $\{ e\otimes e \, , \, e\otimes a \, , \, a\otimes e \, , \, a\otimes a \}$.

The above consideration can be formulated in group theory in a more abstract way.

Let G be a group. Let $g_j \in G$. We consider now the set of all ordered pairs (g_j, g_j). We define

$$(g_j, g_j)(g_k, g_k) := (g_j g_k, g_j g_k).$$

With this definition the set $\{(g_j, g_j)\}$ forms a group with the group multiplication defined above.

2.13 Group Representation Theory

Let V be the vector space $\mathbf{C}^n$. Let $GL(n, \mathbf{C})$ be the set of all invertible $n \times n$ matrices over $\mathbf{C}$. Thus $GL(n, \mathbf{C})$ is a group under matrix multiplication. For a more detailed discussion of group representation theory we refer to Miller [24] and Ludwig and Falter [23].

Definition. An n-dimensional *matrix representation* of G is a homomorphism

$$T : g \longrightarrow T(g)$$

of G into $GL(n, \mathbf{C})$.

As a consequence of this definition we find that

$$T(g_1)T(g_2) = T(g_1 g_2), \qquad T(g)^{-1} = T(g^{-1}), \qquad T(e) = I_n, \qquad g_1, g_2, g \in G$$

where e is the identity element of G and I_n is the unit matrix in $\mathbf{C}^n$.

Example. Consider the group $G = \{ e, a \}$ with $ee = e, ea = a, ae = a, aa = e$. Then

$$e \mapsto 1, \qquad a \mapsto 1$$

is a one-dimensional representation. Another one-dimensional representation is

$$e \mapsto 1, \qquad a \mapsto -1.$$

A two-dimensional representation is given by

$$e \mapsto \begin{pmatrix} 1 & 0 \\ 0 & 1 \end{pmatrix}, \qquad a \mapsto \begin{pmatrix} 0 & 1 \\ 1 & 0 \end{pmatrix}. \qquad \clubsuit$$

Definition. The *character* of $T(g)$ is defined by

$$\chi(g) := \mathrm{tr}(T(g))$$

where tr denotes the trace.

Any group representation T of G with representation space V defines many matrix representations. For, if $\{\mathbf{v}_1, \ldots, \mathbf{v}_n\}$ is a basis of $\mathbf{C}^n$, the matrices $T(g) = (T(g)_{kj})$ defined by

$$T(g)\mathbf{v}_k = \sum_{j=1}^{n} T(g)_{jk}\mathbf{v}_j, \qquad 1 \le k \le n$$

form an n-dimensional matrix representation of G. Every choice of a basis for V yields a new matrix representation of G defined by T. However, any two such matrix representations T, T' are equivalent in the sense that there exists a matrix $S \in GL(n, \mathbf{C})$ such that

$$T'(g) = ST(g)S^{-1}$$

for all $g \in G$. If T, T' correspond to the bases $\{ \mathbf{v}_i \}$, $\{ \mathbf{v}'_i \}$ respectively, then for S we can take the matrix (S_{ji}) defined by

$$\mathbf{v}_i = \sum_{j=1}^{n} S_{ji} \mathbf{v}'_j, \qquad i = 1, \ldots, n.$$

In order to determine all possible representations of a group G it is enough to find one representation T in each equivalence class.

Let T be a finite-dimensional representation of a finite group G acting on the complex vector space $V = \mathbf{C}^n$.

Definition. A subspace W of V is *invariant* under T if

$$T(g)\mathbf{w} \in W$$

for every $g \in G, \mathbf{w} \in W$

If W is invariant under T we can define a representation $T' = T|W$ of G on W by

$$T'(g)\mathbf{w} := T(g)\mathbf{w} \qquad \mathbf{w} \in W.$$

This representation is called the restriction of T to W. If T is a unitary matrix so is T'.

Definition. The representation T is *reducible* if there is a proper subspace W of V which is invariant under T. Otherwise, T is *irreducible.*

A representation is irreducible if the only invariant subspaces of V are $\{\mathbf{0}\}$ and V itself. One-dimensional and zero-dimensional representations are necessarily irreducible.

Suppose T is reducible and W is a proper invariant subspace of V. If $\dim W = k$ and $\dim V = n$ we can find a basis $\mathbf{v}_1, \ldots, \mathbf{v}_n$ for V such that $\mathbf{v}_1, \ldots, \mathbf{v}_k$, $1 \leq k \leq n$, form a basis for W. Then the matrices $T(g)$ with respect to this basis take the form

$$\begin{pmatrix} T'(g) & *** \\ 0 & T''(g) \end{pmatrix}$$

The $k \times k$ matrices $T'(g)$ and the $(n-k) \times (n-k)$ matrices $T''(g)$ separately define matrix representations of G. In particular $T'(g)$ is the matrix of the representation $T'(g)$, with respect to the basis $\mathbf{v}_1, \ldots, \mathbf{v}_k$ of W. Here 0 is the $(n-k)k$ zero matrix.

Every reducible representation can be decomposed into irreducible representations in an almost unique manner. Thus the problem of constructing all representations of G simplifies to the problem of constructing all irreducible representations.

Definition. Let V and V' be vector spaces of dimensions n, m, respectively. Let $\{\mathbf{u}_j\}$, $\{\mathbf{u}'_k\}$ be bases for these spaces. We define $V \otimes V'$, the direct product (also called tensor product) of V and V', as the $n \times m$ dimensional vector space with basis

$$\{\mathbf{u}_j \otimes \mathbf{u}'_k\} \qquad 1 \le j \le n, \qquad 1 \le k \le m.$$

Thus, any $\mathbf{w} \in V \otimes V'$ can be written uniquely in the form

$$\mathbf{w} = \sum_{j=1}^{n} \sum_{k=1}^{m} \alpha_{jk} \mathbf{u}_j \otimes \mathbf{u}'_k.$$

If

$$\mathbf{u} = \sum_{j=1}^{n} \alpha_j \mathbf{u}_j, \qquad \mathbf{u}' = \sum_{k=1}^{m} \beta_k \mathbf{u}'_k$$

we define the vector $\mathbf{u} \otimes \mathbf{u}' \in V \otimes V'$ by

$$\mathbf{u} \otimes \mathbf{u}' := \sum_{j=1}^{n} \sum_{k=1}^{m} \alpha_j \beta_k \mathbf{u}_j \otimes \mathbf{u}'_k.$$

Definition: If T_1 and T_2 are matrix representation of the groups G_1 and G_2, respectively, we can define a matrix representation T of the direct product group $G_1 \times G_2$ on $V_1 \otimes V_2$ by

$$(T(g_1 g_2))(\mathbf{u}_1 \otimes \mathbf{u}_2) := T_1(g_1)\mathbf{u}_1 \otimes T_2(g_2)\mathbf{u}_2$$

where $g_1 \in G_1$, $g_2 \in G_2$, $\mathbf{u}_1 \in V_1$, $\mathbf{u}_2 \in V_2$.

If T_1 is n_1-dimensional and T_2 is n_2-dimensional, then T is $n_1 n_2$ dimensional. Furthermore, we find that the character χ of T is given by

$$\chi(g_1 g_2) = \chi_1(g_1)\chi_2(g_2)$$

where χ_j is the character of T_j. This is due to the fact that $\mathrm{tr}(A \otimes B) \equiv \mathrm{tr}(A)\mathrm{tr}(B)$, where A is an $m \times m$ matrix and B is an $n \times n$ matrix.

Theorem. The matrix representation T is irreducible if and only if T_1 and T_2 are irreducible.

For the proof we refer to the books of Miller [24] and Ludwig and Falter [23].

Exercises. (1) Show that

$$e \mapsto \begin{pmatrix} 1 & 0 & 0 \\ 0 & 1 & 0 \\ 0 & 0 & 1 \end{pmatrix}, \qquad a \mapsto \begin{pmatrix} 0 & 0 & 1 \\ 0 & 1 & 0 \\ 1 & 0 & 0 \end{pmatrix}$$

is a three dimensional representation of the group $G = \{e, a\}$.

(2) Consider the group $G = \{e, a\}$ with $ee = e, ea = a, ae = a, aa = e$. Show that the representations $e \mapsto 1$, $a \mapsto 1$ and $e \mapsto 1, a \mapsto -1$ are irreducible. Show that the representations

$$e \mapsto \begin{pmatrix} 1 & 0 \\ 0 & 1 \end{pmatrix}, \qquad a \mapsto \begin{pmatrix} 1 & 0 \\ 0 & -1 \end{pmatrix}$$

and

$$e \mapsto \begin{pmatrix} 1 & 0 \\ 0 & 1 \end{pmatrix}, \qquad a \mapsto \begin{pmatrix} 0 & 1 \\ 1 & 0 \end{pmatrix}$$

are reducible.

(3) Show that the set

$$\{ +1, -1, +i, -i \}$$

forms a group under multiplication. Is the group Abelian ? Find a two-dimensional representation. Find all irreducible representations. We recall that the number of irreducible representations is equal to the number of conjugacy classes. Thus first find all the conjugacy classes.

(4) Consider the group G_1 of all 2×2 permutation matrices and the group G_2 of all 3×3 permutation matrices. Obviously $V_1 = \mathbf{R}^2$ and $V_2 = \mathbf{R}^3$. Find reducible and irreducible matrix representations of the direct product group $G_1 \times G_2$ on $V_1 \otimes V_2$.

2.14 Inversion of Partitioned Matrices

Partitioned matrices where the elements in each block satisfy some particular pattern, arise widely in systems theory. Block Toeplitz, block Hankel and block circulant matrices play a central role in linear control theory, signal processing and statistics (Davis [10]). Any partitioned matrix, each of whose blocks has the same pattern amongst its elements, is similar to a matrix in which this pattern occurs in block form. This transformation is advantageous for the purpose of inversion. A simple derivation (Barnett [4]) is given using the Kronecker matrix product.

Suppose that A is an $mn \times mn$ matrix partioned into blocks A_{ij}, each of which is $n \times n$ and has the same pattern amongst its elements. We show that inversion of A is equivalent to inverting a matrix B in which the pattern appears amongst the blocks. For example, if each A_{ij} has Toeplitz form, then B is a block Toeplitz matrix, and can be efficiently inverted using a standard algorithm.

Let X be an $N \to M$ matrix having columns $\mathbf{x}_1, \ldots, \mathbf{x}_M$, and define an $MN \times 1$ column vector, $\mathrm{vec}(X)$ by (see sections 1.8 and 2.11)

$$\mathrm{vec}(X) := \begin{pmatrix} \mathbf{x}_1 \\ \mathbf{x}_2 \\ \vdots \\ \mathbf{x}_M \end{pmatrix}.$$

The $MN \times MN$ vec-permutation matrix $P_{N,M}$ is then defined by

$$\mathrm{vec}(X) = P_{N,M}\mathrm{vec}(X^T)$$

where the superscript T denotes transpose. It follows that

$$P_{N,M}P_{M,N} = I_{MN}$$

where I_{MN} is the unit matrix of order MN. The key property of the vec-permutation matrix is that it reverses the order of a Kronecker product $X \otimes Y$ (here Y is $P \times R$), namely

$$P_{N,P}(X \otimes Y)P_{R,M} = Y \otimes X.$$

In the following we assume that $m = 2$, so that

$$A := \begin{pmatrix} A_{11} & A_{12} \\ A_{21} & A_{22} \end{pmatrix}$$

and let each block be an $n \times n$ Toeplitz matrix, i.e.

$$A_{ij} = (a_{rs}^{(ij)}) = (a_{r-s}^{(ij)}), \qquad i,j = 1,2; \qquad r,s = 1,2,\ldots,n$$

The matrix A can be written as

$$\begin{aligned} A &= B_0 \otimes I_n + B_{-1} \otimes K_1 + B_{-2} \otimes K_2 + \cdots + B_{1-n} \otimes K_{n-1} \\ &+ B_1 \otimes K_1^T + B_2 \otimes K_2^T + \cdots + B_{n-1} \otimes K_{n-1}^T \end{aligned}$$

where

$$B_k := \begin{pmatrix} a_k^{(11)} & a_k^{(12)} \\ a_k^{(21)} & a_k^{(22)} \end{pmatrix}$$

for $k = 1-n, 2-n, \ldots, n-2, n-1$ and I_n is the $n \times n$ unit matrix. Here, K_i denotes an $n \times n$ matrix all of whose elements are zero except for a line of ones parallel to, and above, the principal diagonal, from entry $(1, i+l)$ to entry $(n-1, n)$. Applying $P_{N,P}(X \otimes Y)P_{R,M} = Y \otimes X$ with

$$N = M = 2, \qquad P = R = n$$

yields

$$\begin{aligned} I_{2,n} A I_{n,2} &= I_n \otimes B_0 + K_1 \otimes B_{-1} + \cdots + K_{n-1} \otimes B_{1-n} \\ &+ K_1^T \otimes B_1 + \ldots + K_{n-1}^T \otimes B_{n-1} \end{aligned}$$

Thus the right hand side is a $2n \times 2n$ matrix B partitioned into 2×2 blocks

$$B_{rs} = B_{r-s}, \qquad r, s = 1, \ldots, n$$

Therefore, the $2n \times 2n$ matrix A having Toeplitz blocks is similar to this block Toeplitz matrix B. The result can be extended (Barnett [4]).

Chapter 3

Applications

3.1 Spin Matrices

In this section we introduce the Pauli spin matrices and give some application in connection with the Kronecker product. In the first application we calculate the eigenvalues and eigenvectors of the two-point Heisenberg model. In the second appplication we show that the representation of the Clifford algebra can be expressed with the help of the Kronecker product.

The *Pauli spin matrices* are defined by

$$\sigma_x := \begin{pmatrix} 0 & 1 \\ 1 & 0 \end{pmatrix}, \qquad \sigma_y := \begin{pmatrix} 0 & -i \\ i & 0 \end{pmatrix}, \qquad \sigma_z := \begin{pmatrix} 1 & 0 \\ 0 & -1 \end{pmatrix}.$$

Instead of σ_x, σ_y and σ_z we also write sometimes σ_1, σ_2 and σ_3. The matrices σ_+ and σ_- are defined as

$$\sigma_+ := \sigma_x + i\sigma_y \equiv \begin{pmatrix} 0 & 2 \\ 0 & 0 \end{pmatrix}, \qquad \sigma_- := \sigma_x - i\sigma_y \equiv \begin{pmatrix} 0 & 0 \\ 2 & 0 \end{pmatrix}.$$

First we list the properties of the Pauli matrices. We find

$$(\sigma_x)^2 = (\sigma_y)^2 = (\sigma_z)^2 = \begin{pmatrix} 1 & 0 \\ 0 & 1 \end{pmatrix} \equiv I_2$$

and

$$\sigma_x\sigma_y = i\sigma_z \qquad \sigma_y\sigma_z = i\sigma_x \qquad \sigma_z\sigma_x = i\sigma_y\,.$$

The matrices are Hermitian and unitary, respectively. This means

$$(\sigma_x)^* = \sigma_x, \qquad (\sigma_y)^* = \sigma_y, \qquad (\sigma_z)^* = \sigma_z$$

and

$$(\sigma_x)^* = (\sigma_x)^{-1}, \qquad (\sigma_y)^* = (\sigma_y)^{-1}, \qquad (\sigma_z)^* = (\sigma_z)^{-1}.$$

Therefore the eigenvalues of σ_x, σ_y and σ_z are given by $\{+1, \ -1\}$. The normalized eigenvectors of σ_x are

$$\frac{1}{\sqrt{2}}\begin{pmatrix} 1 \\ 1 \end{pmatrix}, \qquad \frac{1}{\sqrt{2}}\begin{pmatrix} 1 \\ -1 \end{pmatrix}.$$

The normalized eigenvectors of σ_y are given by

$$\frac{1}{\sqrt{2}}\begin{pmatrix} 1 \\ i \end{pmatrix}, \qquad \frac{1}{\sqrt{2}}\begin{pmatrix} 1 \\ -i \end{pmatrix}.$$

For σ_z we obviously find the eigenvectors

$$\begin{pmatrix} 1 \\ 0 \end{pmatrix}, \qquad \begin{pmatrix} 0 \\ 1 \end{pmatrix}.$$

The commutator of the Pauli spin matrices gives

$$[\sigma_x, \sigma_y] = 2i\sigma_z, \qquad [\sigma_z, \sigma_x] = 2i\sigma_y, \qquad [\sigma_y, \sigma_z] = 2i\sigma_x.$$

The Pauli spin matrices form a basis of a Lie algebra under the commutator.

The anticommutator of the Pauli spin matrices vanishes, i.e.

$$[\sigma_x, \sigma_y]_+ = 0, \qquad [\sigma_z, \sigma_x]_+ = 0, \qquad [\sigma_y, \sigma_z]_+ = 0.$$

The *spin matrices* are defined by

$$S_x := \frac{1}{2}\sigma_x, \qquad S_y := \frac{1}{2}\sigma_y, \qquad S_z := \frac{1}{2}\sigma_z.$$

Definition. Let $j = 1, 2, \ldots, N$. We define

$$\sigma_{\alpha,j} := I \otimes \ldots \otimes I \otimes \sigma_\alpha \otimes I \otimes \ldots \otimes I$$

where I is the 2×2 unit matrix, $\alpha = x, y, z$ and σ_α is the α-th Pauli matrix in the j-th location. Thus $\sigma_{\alpha,j}$ is a $2^N \times 2^N$ matrix.

Analogously, we define

$$S_{\alpha,j} := I \otimes \ldots \otimes I \otimes S_\alpha \otimes I \otimes \ldots \otimes I.$$

In the following we set

$$\mathbf{S}_j := (S_{x,j}, S_{y,j}, S_{z,j})^T.$$

Example. We calculate the eigenvalues and eigenvectors for the two-point *Heisenberg model.* The model is given by the Hamilton operator

$$\hat{H} = J\sum_{j=1}^{2} \mathbf{S}_j \cdot \mathbf{S}_{j+1}$$

where J is the so-called exchange constant ($J > 0$ or $J < 0$) and $\cdot$ denotes the scalar product. We impose *cyclic boundary conditions*, i.e.

$$\mathbf{S}_3 \equiv \mathbf{S}_1.$$

It follows that

$$\hat{H} = J(\mathbf{S}_1 \cdot \mathbf{S}_2 + \mathbf{S}_2 \cdot \mathbf{S}_3) \equiv J(\mathbf{S}_1 \cdot \mathbf{S}_2 + \mathbf{S}_2 \cdot \mathbf{S}_1).$$

Therefore

$$\hat{H} = J(S_{x,1}S_{x,2} + S_{y,1}S_{y,2} + S_{z,1}S_{z,2} + S_{x,2}S_{x,1} + S_{y,2}S_{y,1} + S_{z,2}S_{z,1}).$$

Since

$$S_{x,1} = S_x \otimes I, \qquad S_{x,2} = I \otimes S_x$$

etc. where I is the 2×2 unit matrix, it follows that

$$\begin{aligned} \hat{H} &= J[(S_x \otimes I)(I \otimes S_x) + (S_y \otimes I)(I \otimes S_y) + (S_z \otimes I)(I \otimes S_z) \\ &\quad +(I \otimes S_x)(S_x \otimes I) + (I \otimes S_y)(S_y \otimes I) + (I \otimes S_z)(S_z \otimes I)]. \end{aligned}$$

We find

$$\hat{H} = J[(S_x \otimes S_x) + (S_y \otimes S_y) + (S_z \otimes S_z) + (S_x \otimes S_x) + (S_y \otimes S_y) + (S_z \otimes S_z)].$$

Therefore

$$\hat{H} = 2J[(S_x \otimes S_x) + (S_y \otimes S_y) + (S_z \otimes S_z)].$$

Since

$$S_x := \frac{1}{2}\sigma_x, \qquad S_y := \frac{1}{2}\sigma_y, \qquad S_z := \frac{1}{2}\sigma_z$$

we obtain

$$S_x \otimes S_x = \frac{1}{4}\begin{pmatrix} 0 & 1 \\ 1 & 0 \end{pmatrix} \otimes \begin{pmatrix} 0 & 1 \\ 1 & 0 \end{pmatrix} = \frac{1}{4}\begin{pmatrix} 0 & 0 & 0 & 1 \\ 0 & 0 & 1 & 0 \\ 0 & 1 & 0 & 0 \\ 1 & 0 & 0 & 0 \end{pmatrix}$$

etc.. Then the Hamilton operator $\hat{H}$ is given by the 4×4 symmetric matrix

$$\hat{H} = \frac{J}{2}\begin{pmatrix} 1 & 0 & 0 & 0 \\ 0 & -1 & 2 & 0 \\ 0 & 2 & -1 & 0 \\ 0 & 0 & 0 & 1 \end{pmatrix} = \frac{J}{2}\left[(1) \oplus \begin{pmatrix} -1 & 2 \\ 2 & -1 \end{pmatrix} \oplus (1)\right]$$

where $\oplus$ denotes the direct sum. The eigenvalues and eigenvectors can now easily be calculated. We define

$$| \uparrow \rangle := \begin{pmatrix} 1 \\ 0 \end{pmatrix} \qquad \text{spin up}$$

and

$$| \downarrow \rangle := \begin{pmatrix} 0 \\ 1 \end{pmatrix} \qquad \text{spin down.}$$

Then we define

$$| \uparrow\uparrow \rangle := | \uparrow \rangle \otimes | \uparrow \rangle, \qquad | \uparrow\downarrow \rangle := | \uparrow \rangle \otimes | \downarrow \rangle, \qquad | \downarrow\uparrow \rangle := | \downarrow \rangle \otimes | \uparrow \rangle, \qquad | \downarrow\downarrow \rangle := | \downarrow \rangle \otimes | \downarrow \rangle.$$

Consequently,

$$| \uparrow\uparrow \rangle = \begin{pmatrix} 1 \\ 0 \\ 0 \\ 0 \end{pmatrix}, \quad | \uparrow\downarrow \rangle = \begin{pmatrix} 0 \\ 1 \\ 0 \\ 0 \end{pmatrix}, \quad | \downarrow\uparrow \rangle = \begin{pmatrix} 0 \\ 0 \\ 1 \\ 0 \end{pmatrix}, \quad | \downarrow\downarrow \rangle = \begin{pmatrix} 0 \\ 0 \\ 0 \\ 1 \end{pmatrix}.$$

One sees at once that $| \uparrow\uparrow \rangle$ and $| \downarrow\downarrow \rangle$ are eigenvectors of the Hamilton operator with eigenvalues $J/2$ and $J/2$, respectively. This means the eigenvalue $J/2$ is degenerate. The eigenvalues of the matrix

$$\frac{J}{2} \begin{pmatrix} -1 & 2 \\ 2 & -1 \end{pmatrix}$$

are given by

$$\left\{ \frac{J}{2}, \quad -\frac{3J}{2} \right\}.$$

The corresponding eigenvectors are given by

$$\frac{1}{2}(| \uparrow\downarrow \rangle + | \downarrow\uparrow \rangle), \qquad \frac{1}{2}(| \uparrow\downarrow \rangle - | \downarrow\uparrow \rangle).$$

Thus the eigenvalue $J/2$ is three times degenerate. If $J > 0$, then $-3J/2$ is the ground state energy and the ground state is non-degenerate. If $J < 0$, then $J/2$ is the ground state energy. ♣

Example. Within the representation of the *Clifford algebra* the Pauli spin matrices and Kronecker product play the central role (Steiner [40]). For every pair of non-negative integers (p, q) we denote by $C(p, q)$ the Clifford algebra of the non-degenerate real bilinear symmetric from of rank $p+q$ and signature $p-q$. This is a 2^{p+q}-dimensional real algebra with generators $e_1, \ldots, e_{p+q}$ satisfying the following relations

$$
\begin{aligned}
e_i^2 &= 1 && \text{for } \ i = 1, \ldots, p \\
e_i^2 &= -1 && \text{for } \ i = p+1, \ldots, p+q \\
e_i e_j &= -e_j e_i && \text{for } \ i \neq j.
\end{aligned}
$$

We can now give a description of these representations. We express the image of the generators e_i of the Clifford algebras as a Kronecker product of real 2×2 matrices

$$
\sigma_0 = \begin{pmatrix} 1 & 0 \\ 0 & 1 \end{pmatrix}, \quad \sigma_1 = \begin{pmatrix} 0 & 1 \\ 1 & 0 \end{pmatrix}, \quad \rho_2 = \begin{pmatrix} 0 & -1 \\ 1 & 0 \end{pmatrix}, \quad \sigma_3 = \begin{pmatrix} 1 & 0 \\ 0 & -1 \end{pmatrix}
$$

where

$$\rho_2 := -i\sigma_2.$$

We find that

$$\rho_2^2 = -I$$

where I is the 2×2 identity matrix.

We obtain at once the irreducible representation of $C(1,1)$:

$$
\begin{aligned}
\alpha : C(1,1) &\rightarrow M_2(\mathbf{R}) \\
f_1 &\mapsto \sigma_1 \\
f_2 &\mapsto \rho_2
\end{aligned}
$$

where $M_2(\mathbf{R})$ is the four-dimensional vector space of the 2×2 matrices over $\mathbf{R}$ and f_1, f_2 are the two generators of the presentation given above.

Owing to the modulo-8 periodicity in the structure of the algebras, the basic representations are those of $C(0,8)$ and $C(8,0)$. We write $g_1, \ldots, g_8$ and $h_1, \ldots, h_8$, respectively. The quantities $g_1, \ldots, g_8$ and $h_1, \ldots, h_8$ are 16×16 matrices over the real numbers $\mathbf{R}$ and given by the Kronecker product of 2×2 matrices.

$$\begin{array}{rcl}
\beta & : & C(0,8) \to M_{16}(\mathbf{R}) \\
g_1 & \mapsto & \rho_2 \otimes I \otimes I \otimes I \\
g_2 & \mapsto & \sigma_3 \otimes \rho_2 \otimes I \otimes I \\
g_3 & \mapsto & \sigma_1 \otimes \rho_2 \otimes \sigma_1 \otimes I \\
g_4 & \mapsto & \sigma_3 \otimes \sigma_3 \otimes \rho_2 \otimes I \\
g_5 & \mapsto & \sigma_1 \otimes I \otimes \rho_2 \otimes \sigma_1 \\
g_6 & \mapsto & \sigma_3 \otimes \sigma_1 \otimes \rho_2 \otimes \sigma_1 \\
g_7 & \mapsto & \sigma_1 \otimes \rho_2 \otimes \sigma_3 \otimes \sigma_1 \\
g_8 & \mapsto & \sigma_3 \otimes \sigma_3 \otimes \sigma_3 \otimes \rho_2
\end{array}$$

and

$$\begin{array}{rcl}
\tau & : & C(8,0) \to M_{16}(\mathbf{R}) \\
h_1 & \mapsto & \sigma_1 \otimes I \otimes I \otimes I \\
h_2 & \mapsto & \sigma_3 \otimes \sigma_1 \otimes I \otimes I \\
h_3 & \mapsto & \sigma_3 \otimes \sigma_3 \otimes \sigma_1 \otimes I \\
h_4 & \mapsto & \rho_2 \otimes \sigma_1 \otimes \rho_2 \otimes I \\
h_5 & \mapsto & \sigma_3 \otimes \sigma_3 \otimes \sigma_4 \otimes \sigma_1 \\
h_6 & \mapsto & \rho_2 \otimes I \otimes \sigma_1 \otimes \rho_2 \\
h_7 & \mapsto & \sigma_3 \otimes \rho_2 \otimes \sigma_1 \otimes \rho_2 \\
h_8 & \mapsto & \rho_2 \otimes \sigma_1 \otimes \sigma_3 \otimes \rho_2
\end{array}$$

We have

$$g_j^2 \mapsto -I_{16}$$

for $j = 1, 2, \ldots, 8$. Analogously, we have

$$h_j^2 \mapsto I_{16}$$

for $j = 1, 2, \ldots, 8$.

The representations of $C(0,n)$ and $C(n,0)$ with $n < 8$ can easily be found from those of $C(0,8)$ and $C(8,0)$, respectively. One has to consider the first n generators and take the first factors in their images, in order to obtain representations of dimension $d(0,n)$ and $d(n,0)$, respectively. ♣

Exercises. (1) Show that any 2×2 matrix can be written as a linear combination of the 2×2 unit matrix and the Pauli matrices.

(2) Show that σ_x, σ_y and σ_z form a basis of a Lie algebra under the commutator.

(3) Let I be the 2×2 unit matrix. Do the matrices

$$\{ \sigma_x \otimes I,\ I \otimes \sigma_x,\ \sigma_y \otimes I,\ I \otimes \sigma_y,\ \sigma_z \otimes I, I \otimes \sigma_z \}$$

form a basis of a Lie algebra under the commutator ?

(4) Show that

$$\exp(\epsilon i \sigma_y) = \begin{pmatrix} \cos(\epsilon) & \sin(\epsilon) \\ -\sin(\epsilon) & \cos(\epsilon) \end{pmatrix}$$

where ϵ is a parameter ($\epsilon \in \mathbf{R}$). Find

$$\exp(\epsilon i \sigma_y \otimes \sigma_z) .$$

(5) Show that

$$\exp(\sigma_x + \sigma_y) \neq \exp(\sigma_x) \exp(\sigma_y).$$

(6) Show that the matrices

$$\sigma_x \otimes I_2, \qquad I_2 \otimes \sigma_y, \qquad \sigma_x \otimes \sigma_y$$

commute with each other. Then show that

$$\exp(\sigma_x \otimes I_2 + I_2 \otimes \sigma_y + \sigma_x \otimes \sigma_y) \equiv \exp(\sigma_x \otimes I_2) \exp(I_2 \otimes \sigma_y) \exp(\sigma_x \otimes \sigma_y).$$

(7) Find all the eigenvalues of the matrices

$$\sigma_x \otimes \sigma_x \otimes \sigma_x \otimes \sigma_x$$

$$\sigma_y \otimes \sigma_y \otimes \sigma_y \otimes \sigma_y$$

$$\sigma_z \otimes \sigma_z \otimes \sigma_z \otimes \sigma_z .$$

(8) Let

$$\hat{H} = J \sum_{j=1}^{3} \mathbf{S}_j \cdot \mathbf{S}_{j+1}$$

with $\mathbf{S}_4 = \mathbf{S}_1$, i. e. we have cyclic boundary conditions. Here

$$S_{x,1} = S_x \otimes I \otimes I, \qquad S_{x,2} = I \otimes S_x \otimes I, \qquad S_{x,3} = I \otimes I \otimes S_x$$

and I is the 2×2 unit matrix. Find the eigenvalues of $\hat{H}$.

(9) Let

$$\hat{H} = J(\mathbf{S}_1 \cdot \mathbf{S}_2 + \mathbf{S}_2 \cdot \mathbf{S}_3).$$

Find the eigenvalues and eigenvectors of $\hat{H}$. Here we have open end boundary conditions.

(10) The following 4×4 matrices

$$A_1 = \begin{pmatrix} 0 & 0 & 0 & 1 \\ 0 & 0 & 1 & 0 \\ 0 & 1 & 0 & 0 \\ 1 & 0 & 0 & 0 \end{pmatrix}$$

$$A_2 = \begin{pmatrix} 0 & 0 & 0 & -i \\ 0 & 0 & i & 0 \\ 0 & -i & 0 & 0 \\ i & 0 & 0 & 0 \end{pmatrix}$$

$$A_3 = \begin{pmatrix} 0 & 0 & 1 & 0 \\ 0 & 0 & 0 & -1 \\ 1 & 0 & 0 & 0 \\ 0 & -1 & 0 & 0 \end{pmatrix}$$

play an important role in the *Dirac theory* of the electron. Show that the matrices A_1, A_2 and A_3 can be written as Kronecker products of the Pauli spin matrices.

(11) For describing spin-1 particles (for example the photon) the following matrices are important

$$A_1 = \begin{pmatrix} 0 & 0 & 0 \\ 0 & 0 & -i \\ 0 & i & 0 \end{pmatrix}, \qquad A_2 = \begin{pmatrix} 0 & 0 & i \\ 0 & 0 & 0 \\ -i & 0 & 0 \end{pmatrix}, \qquad A_3 = \begin{pmatrix} 0 & -i & 0 \\ i & 0 & 0 \\ 0 & 0 & 0 \end{pmatrix}.$$

Find the eigenvalues and eigenvectors of A_1, A_2 and A_3. Find the eigenvalues and eigenvectors of $A_1 \otimes A_1$, $A_2 \otimes A_2$ and $A_3 \otimes A_3$.

3.2 One-Dimensional Ising Model

The one-dimensional *Ising model* with cyclic boundary conditions is given by

$$\hat{H} = -J \sum_{j=1}^{N} \sigma_{z,j} \sigma_{z,j+1}$$

where

$$\sigma_{z,N+1} \equiv \sigma_{z,1}.$$

This means we impose cyclic boundary conditions. Obviously $\hat{H}$ is a $2^N \times 2^N$ diagonal matrix. Here N is the number of lattice sites and J is a nonzero real constant, the so-called exchange constant. We calculate the *partition function* $Z(\beta J)$ which is defined by

$$Z_N(\beta J) := \mathrm{tr} e^{-\beta \hat{H}} \equiv \mathrm{tr} \exp\left(\beta J \sum_{j=1}^{N} \sigma_{z,j} \sigma_{z,j+1} \right)$$

where $\beta = 1/(kT)$. Here k denotes the Boltzmann constant and T is the absolute temperature. From the partition function Z_N we obtain the *Helmholtz free energy* per lattice site

$$\frac{F(\beta J)}{N} := -\frac{1}{N\beta} \ln Z_N(\beta J).$$

We first consider the case with four lattice points, i.e. $N = 4$. Then we extend to arbitrary N.

Since

$$\sigma_{z,j} = \overbrace{I \otimes \ldots \otimes I \otimes \underset{\substack{\uparrow \\ j-\text{th place}}}{\sigma_z} \otimes I \otimes \ldots \otimes I}^{\text{N-factors}}$$

we have for $N = 4$

$$\begin{aligned} \sigma_{z,1} &= \sigma_z \otimes I \otimes I \otimes I \\ \sigma_{z,2} &= I \otimes \sigma_z \otimes I \otimes I \\ \sigma_{z,3} &= I \otimes I \otimes \sigma_z \otimes I \\ \sigma_{z,4} &= I \otimes I \otimes I \otimes \sigma_z. \end{aligned}$$

We set

$$X_{12} := \sigma_{z,1}\sigma_{z,2}, \qquad X_{23} := \sigma_{z,2}\sigma_{z,3}, \qquad X_{34} := \sigma_{z,3}\sigma_{z,4}, \qquad X_{41} := \sigma_{z,4}\sigma_{z,1}.$$

Consequently

$$\begin{aligned} X_{12} &\equiv \sigma_z \otimes \sigma_z \otimes I \otimes I \\ X_{23} &\equiv I \otimes \sigma_z \otimes \sigma_z \otimes I \\ X_{34} &\equiv I \otimes I \otimes \sigma_z \otimes \sigma_z \\ X_{41} &\equiv \sigma_z \otimes I \otimes I \otimes \sigma_z. \end{aligned}$$

Let

$$M := \{ X_{12}, X_{23}, X_{34}, X_{41} \}.$$

Thus the elements of M are $2^4 \times 2^4$ matrices. We recall the identities

$$\operatorname{tr}(A + B) \equiv \operatorname{tr}A + \operatorname{tr}B$$

$$\operatorname{tr}(A \otimes B) \equiv (\operatorname{tr}A)(\operatorname{tr}B)$$

where A and B are $n \times n$ matrices.

We need the following lemmata.

Lemma. Let X be a real $n \times n$ matrix such that $X^2 = I_n$. Let $\epsilon \in \mathbf{R}$. Then

$$\exp(\epsilon X) \equiv I_n \cosh \epsilon + X \sinh \epsilon.$$

Remark. The matrices given above are examples of such matrices.

Lemma. Let X and Y be arbitrary elements of M. Then

$$[X, Y] = 0.$$

Lemma. Let X, Y and Z be arbitrary elements of M. Then

$$\begin{aligned} \operatorname{tr}X &= 0 \\ \operatorname{tr}(XY) &= 0 \quad \text{if} \quad X \neq Y \\ \operatorname{tr}(XYZ) &= 0 \quad \text{if} \quad X \neq Y. \end{aligned}$$

Lemma. Let $X_{12}, X_{23}, X_{34}, X_{41}$ be the matrices given above. Then

$$\begin{aligned} X_{12}X_{23}X_{34}X_{41} &= I_{16} \\ \operatorname{tr}(X_{12}X_{23}X_{34}X_{41}) &= 2^4 = 16. \end{aligned}$$

Theorem. Let

$$K \equiv \epsilon(X_{12} + X_{23} + X_{34} + X_{41}).$$

Then

$$\operatorname{tr}\exp(K) \equiv \operatorname{tr}(I_{16}\cosh^4\epsilon + I_{16}\sinh^4\epsilon).$$

Proof. Using the lemmata given above we find

$$\begin{aligned}
\operatorname{tr}\exp(K) &\equiv \operatorname{tr}[\exp(\epsilon X_{12})\exp(\epsilon X_{23})\exp(\epsilon X_{34})\exp(\epsilon X_{41})] \\
&\equiv \operatorname{tr}[(I_{16}\cosh\epsilon + X_{12}\sinh\epsilon)(I_{16}\cosh\epsilon + X_{23}\sinh\epsilon) \\
&\quad \times(I_{16}\cosh\epsilon + X_{34}\sinh\epsilon)(I_{16}\cosh\epsilon + X_{41}\sinh\epsilon)] \\
&\equiv \operatorname{tr}\left[I_{16}\cosh^4\epsilon + I_{16}\sinh^4\epsilon\right]. \qquad \spadesuit
\end{aligned}$$

As a consequence we obtain

$$\operatorname{tr}(I_{16}\cosh^4\epsilon + I_{16}\sinh^4\epsilon) \equiv 2^4(\cosh^4\epsilon + \sinh^4\epsilon).$$

Consequently the partition function is given by

$$Z_4(\beta J) = 2^4(\cosh^4(\beta J) + \sinh^4(\beta J)).$$

From the partition function we find the Helmholtz free energy.

Next we consider the case with arbitrary N.

Theorem. Let

$$\begin{aligned}
X_{12} &\equiv \overbrace{\sigma_z \otimes \sigma_z \otimes I \otimes \cdots \otimes I}^{N\text{-factors}} \\
X_{23} &\equiv I \otimes \sigma_z \otimes \sigma_z \otimes \cdots \otimes I \\
&\vdots \\
X_{N-1,N} &\equiv I \otimes I \otimes \cdots \otimes I \otimes \sigma_z \otimes \sigma_z \\
X_{N1} &\equiv \sigma_z \otimes I \otimes \cdots \otimes I \otimes \sigma_z
\end{aligned}$$

where I is the unit 2×2 matrix. Let

$$K := \epsilon(X_{12} + X_{23} + \cdots + X_{N-1,N} + X_{N1})$$

where $\epsilon \in \mathbf{R}$. Then

$$\operatorname{tr}\exp(K) \equiv \operatorname{tr}(I_{2^N}\cosh^N\epsilon + I_{2^N}\sinh^N\epsilon)$$

where I_{2^N} is the unit $2^N \times 2^N$ matrix.

As a consequence we obtain

$$\mathrm{tr}\exp(K) \equiv 2^N \cosh^N \epsilon + 2^N \sinh^N \epsilon \equiv 2^N(\cosh^N \epsilon + \sinh^N \epsilon).$$

Remark 1. The lemmata described above can easily be extended to arbitrary N. Consequently, the proof of this theorem can be performed in the same way as the proof of the theorem for the case $N = 4$.

Remark 2. When we take the Pauli matrices

$$\sigma_x = \begin{pmatrix} 0 & 1 \\ 1 & 0 \end{pmatrix}$$

or

$$\sigma_y = \begin{pmatrix} 0 & -i \\ i & 0 \end{pmatrix}$$

instead of σ_z, we obtain the same result for the partition function, because

$$\sigma_x^2 = \sigma_y^2 = I.$$

Remark 3. The $2^N \times 2^N$ matrix K describes the one-dimensional Ising model with cyclic boundary conditions. The traditional approach is as follows: in one dimension the Ising model is given by

$$H_N = -J \sum_{j=1}^{N} \sigma_j \sigma_{j+1}$$

where

$$\sigma_{N+1} = \sigma_1$$

(cyclic boundary conditions) and

$$\sigma_j \in \{1, -1\}.$$

Here N is the number of lattice sites, and J $(J > 0)$ is the real constant. The partition function is

$$Z_N(\beta J) = \sum_{\sigma_1, \ldots, \sigma_N \in \Sigma} \exp\left(\beta J \sum_{j=1}^{N} \sigma_j \sigma_{j+1}\right)$$

where the first sum runs over all possible configurations Σ. There are 2^N possible configurations. Hence, we must sum over all possible configurations, namely 2^N, instead of calculating the trace of a $2^N \times 2^N$ matrix.

Exercises. (1) Let

$$\hat{H} = -J \sum_{j=1}^{4} \sigma_{z,j} \sigma_{z,j+1}$$

and $\sigma_{z,5} = 0$. This is the open end boundary condition. Find the partition function $Z(\beta)$. Find the Helmholtz free energy.

(2) Find the eigenvalues of

$$\hat{H} = -J(\sigma_{z,1}\sigma_{z,2} + \sigma_{z,2}\sigma_{z,3}).$$

(3) Find the eigenvalues of

$$\hat{H} = -J(\sigma_{z,1}\sigma_{z,2} + \sigma_{z,2}\sigma_{z,3}) + B(\sigma_{z,1} + \sigma_{z,2} + \sigma_{z,3})$$

where B is a positive constant (the magnetic field).

(4) Given n lattice sites. Each lattice site can either be occupied by a spin up or by a spin down. Show that the number of possible configurations is

$$2^n .$$

Discuss the case where a lattice site can be occupied in three different ways, for example spin up, spin down and spin horizontal.

3.3 Fermi Systems

From statistical mechanics it is known that all equilibrium thermodynamic quantities of interest for a many body system of interacting Fermi particles can be determined from the *grand partition function* [32]

$$Z(\beta) := \operatorname{tr}\exp(-\beta(\hat{H} - \mu\hat{N}_e)).$$

Here $\hat{H}$ is the Hamilton operator describing the system. Throughout, we consider fermi systems in the occupation number formalism. $\hat{N}_e$ is the total number operator, μ the chemical potential, and β the inverse temperature. In general, the trace cannot be evaluated. However, often, the Hamilton operator consists of two terms, namely

$$\hat{H} = \hat{H}_0 + \hat{H}_1$$

where $\hat{H}_0$ is so chosen that the properties described by $\hat{H}_0$ alone are well-known. In most cases stated in the literature, the operator $\hat{H}_0 - \mu\hat{N}_e$ has the form

$$\hat{H}_0 - \mu\hat{N}_e = \sum_{k\sigma}(\varepsilon(k) - \mu)c_{k\sigma}^\dagger c_{k\sigma}$$

where $\varepsilon(k)$ is the one-particle energy, and σ denotes the spin. We investigate a lattice system and therefore k runs over the first Brioullin zone. Here we are able to calculate the trace exactly.

The requirements of the *Pauli principle* are satisfied if the Fermion operators

$$\{\, c_{k\sigma}^\dagger, c_{j\sigma} \,:\, k, j = 1, 2, \ldots, N : \sigma \in \{\uparrow, \downarrow\}\,\}$$

satisfy the anticommutation relations

$$\begin{aligned} [c_{k\sigma}^\dagger, c_{j\sigma'}]_+ &= \delta_{kj}\delta_{\sigma\sigma'}I \\ [c_{k\sigma}^\dagger, c_{j\sigma'}^\dagger]_+ &= [c_{k\sigma}, c_{j\sigma'}]_+ = 0 \end{aligned}$$

for all $k, j = 1, 2, \ldots, N$ and $\sigma, \sigma' \in \{\uparrow, \downarrow\}$. Here I denotes the unit operator.

We show that the trace calculation of $\hat{H}_0 - \mu\hat{N}_e$ can easily be performed with the aid of matrix calculation and the Kronecker product of matrices ([32], [33], [34], [42]).

First we recall that

$$\begin{aligned} &\operatorname{tr}\exp[(A_1 \otimes I \otimes \ldots \otimes I) + (I \otimes A_2 \otimes I \ldots \otimes I) + (I \otimes I \otimes \ldots \otimes A_N)] \\ &= (\operatorname{tr}\exp A_1)(\operatorname{tr}\exp A_2)\ldots(\operatorname{tr}\exp A_N) = \prod_{k=1}^{N}\operatorname{tr}\exp(A_k). \end{aligned}$$

This formula is the main result. It remains to show that the Hamilton operator $\hat{H}_0 - \mu\hat{N}_e$ can be brought into the form

$$(A_1 \otimes I \otimes \ldots \otimes I) + (I \otimes A_2 \otimes I \ldots \otimes I) + \ldots + (I \otimes I \otimes \ldots \otimes A_N)$$

where I is the unit matrix.

Therefore the matrix representation for the Fermi operators will be given. For the sake of simplicity we first consider spinless Fermions. Then the case including the spin is described.

The requirements of the *Pauli principle* are satisfied if the spinless Fermion operators

$$\{ c_k^\dagger, c_j \, : \, k, j = 1, 2, \ldots, N \}$$

satisfy the anticommutation relations

$$\begin{aligned} [c_k^\dagger, c_j]_+ &= \delta_{kj} I \\ [c_k^\dagger, c_j^\dagger]_+ &= [c_k, c_j]_+ = 0 \end{aligned}$$

for all $k, j = 1, 2, \ldots, N$.

Let us now give a faithful matrix representation of the operators. First we discuss the case $N = 1$. A basis is given by

$$\{ c^\dagger|0\rangle, \quad |0\rangle \}$$

and the corresponding dual one by

$$\{ \langle 0|c, \quad \langle 0| \}$$

with

$$\begin{aligned} \langle 0|0\rangle &= 1 \\ \langle 0|cc^\dagger|0\rangle &= 1 \\ \langle 0|c|0\rangle &= 0 \\ \langle 0|c^\dagger|0\rangle &= 0 \end{aligned}$$

where the vector space under consideration is two-dimensional i.e. $\mathbf{C}^2$. Consequently, $c^\dagger$ and c have the faithful matrix representation

$$c^\dagger = \begin{pmatrix} 0 & 1 \\ 0 & 0 \end{pmatrix} = \frac{1}{2}(\sigma_x + i\sigma_y) = \frac{1}{2}\sigma_+$$

$$c = \begin{pmatrix} 0 & 0 \\ 1 & 0 \end{pmatrix} = \frac{1}{2}(\sigma_x - i\sigma_y) = \frac{1}{2}\sigma_- .$$

The *state vectors* are

$$c^\dagger|0\rangle = \begin{pmatrix} 1 \\ 0 \end{pmatrix}, \qquad |0\rangle = \begin{pmatrix} 0 \\ 1 \end{pmatrix}.$$

Here σ_x, σ_y, σ_z, σ_+, σ_- are the usual Pauli matrices.

The number operator $\hat{n}$ defined by

$$\hat{n} := c^\dagger c$$

becomes

$$\hat{n} := c^\dagger c = \begin{pmatrix} 1 & 0 \\ 0 & 0 \end{pmatrix}$$

with the eigenvalues 0 and 1.

The extension to the case $N > 1$ leads to

$$\begin{aligned} c_k^\dagger &= \overbrace{\sigma_z \otimes \sigma_z \otimes \ldots \otimes \sigma_z \otimes \left(\frac{1}{2}\sigma_+\right) \otimes I \otimes I \otimes \ldots \otimes I}^{N\text{-times}} \\ c_k &= \sigma_z \otimes \sigma_z \otimes \ldots \otimes \sigma_z \otimes \underbrace{\left(\frac{1}{2}\sigma_-\right)}_{k\text{-th place}} \otimes I \otimes I \otimes \ldots \otimes I \end{aligned}$$

where I is the 2×2 unit matrix. One can easily calculate that the anticommutation relations are fulfilled. The number operator

$$\hat{n}_k := c_k^\dagger c_k$$

with quantum number k is found to be

$$\hat{n}_k = c_k^\dagger c_k = I \otimes \ldots I \otimes \begin{pmatrix} 1 & 0 \\ 0 & 0 \end{pmatrix} \otimes I \otimes \ldots \otimes I.$$

Finally, one has

$$\begin{aligned} \sum_{k=1}^{N} \lambda_k \hat{n}_k &= \begin{pmatrix} \lambda_1 & 0 \\ 0 & 0 \end{pmatrix} \otimes I \otimes \ldots \otimes I + I \otimes \begin{pmatrix} \lambda_2 & 0 \\ 0 & 0 \end{pmatrix} \otimes I \otimes \ldots \otimes I \\ &+ \ldots + I \otimes \ldots \otimes I \otimes \begin{pmatrix} \lambda_N & 0 \\ 0 & 0 \end{pmatrix} \end{aligned}$$

where $\lambda_k \in \mathbf{R}$. The total number operator is defined as

$$\hat{N}_e := \sum_{k=1}^{N} \hat{n}_k \equiv \sum_{k=1}^{N} c_k^\dagger c_k.$$

The underlying vector space is given by $\mathbf{C}^{2^N}$. Thus the determination of the trace

$$\operatorname{tr} \exp\left(\sum_{k=1}^{N} \lambda_k \hat{n}_k\right)$$

reduces to the trace calculation in a subspace. In occupation number formalism this subspace is given by the basis

$$\left\{ c^\dagger|0\rangle, \quad |0\rangle \right\}$$

and in matrix calculation it is given by the basis

$$\left\{ \begin{pmatrix} 1 \\ 0 \end{pmatrix}, \begin{pmatrix} 0 \\ 1 \end{pmatrix} \right\}.$$

Consequently, we find that

$$\operatorname{tr} \exp\left(\sum_{k=1}^{N} \lambda_k \hat{n}_k\right) = \prod_{k=1}^{N} \operatorname{tr} \exp\begin{pmatrix} \lambda_k & 0 \\ 0 & 0 \end{pmatrix} = \prod_{k=1}^{N} \operatorname{tr}\begin{pmatrix} e^{\lambda_k} & 0 \\ 0 & 1 \end{pmatrix} = \prod_{k=1}^{N} (e^{\lambda_k} + 1).$$

Example. As an example for an approximative solution we consider the *Hubbard model.* The Hamilton operator can be written in *Bloch representation* as

$$\hat{H} = \sum_{k\sigma} \varepsilon(k) c_{k\sigma}^\dagger c_{k\sigma} + \sum_{k_1 k_2 k_3 k_4} \delta(k_1 - k_2 + k_3 - k_4) c_{k_1\uparrow}^\dagger c_{k_2\uparrow} c_{k_3\downarrow}^\dagger c_{k_4\downarrow}$$

where k runs over the first Brioullin zone. As an approximative method we use the following inequality, which is sometimes called *Bogolyubov inequality,*

$$\Omega \le \operatorname{tr}(\hat{H} - \mu \hat{N}_e) W_t + \frac{1}{\beta} \operatorname{tr} W_t \ln W_t$$

for the grand potential Ω. W_t is the so-called trial density matrix. Let us consider the case where

$$W_t = \frac{\exp(-\beta \sum_k (E_\uparrow(k) c_{k\uparrow}^\dagger c_{k\uparrow} + E_\downarrow(k) c_{k\downarrow}^\dagger c_{k\downarrow}))}{\operatorname{tr} \exp(-\beta \sum_k (E_\uparrow(k) c_{k\uparrow}^\dagger c_{k\uparrow} + E_\downarrow(k) c_{k\downarrow}^\dagger c_{k\downarrow}))}.$$

Both $E_\uparrow(k)$ and $E_\downarrow(k)$ play the role of real variation parameters.

For example, we have to calculate traces such as

$$\mathrm{tr}(c^\dagger_{k\uparrow}c_{k\uparrow}W_t)$$

$$\sum_{k_1k_2k_3k_4}\delta(k_1-k_2+k_3-k_4)\mathrm{tr}(c^\dagger_{k_1\uparrow}c_{k_2\uparrow}c^\dagger_{k_3\downarrow}c_{k_4\downarrow}W_t).$$

Now we show how, within the framework of the developed calculation, the trace can be evaluated. Since we have included the spin, the matrix representation of the Fermi operators

$$\{\, c^\dagger_{k\sigma},\ c_{k\sigma} : k=1,\ldots,N;\ \sigma=\uparrow,\ \downarrow\}$$

becomes

$$c^\dagger_{k\uparrow} = \underbrace{\sigma_z\otimes\ldots\otimes\sigma_z\otimes\overset{k\text{-th place}}{\left(\frac{1}{2}\sigma_+\right)}\otimes I\otimes\ldots\otimes I}_{2N\text{ times}}$$

$$c^\dagger_{k\downarrow} = \sigma_z\otimes\ldots\otimes\sigma_z\otimes\underset{(k+N)\text{-th place}}{\left(\frac{1}{2}\sigma_+\right)}\otimes I\otimes\ldots\otimes I.$$

Using

$$\exp\begin{pmatrix}\lambda & 0\\ 0 & 0\end{pmatrix}\equiv\begin{pmatrix}\exp(\lambda) & 0\\ 0 & 1\end{pmatrix}$$

it follows that

$$\exp\left(\sum_k E_\uparrow(k)c^\dagger_{k\uparrow}c_{k\uparrow}+\sum_k E_\downarrow(k)c^\dagger_{k\downarrow}c_{k\downarrow}\right)$$

$$\equiv\begin{pmatrix}e^{E_\uparrow(1)} & 0\\ 0 & 1\end{pmatrix}\otimes\ldots\otimes\begin{pmatrix}e^{E_\uparrow(N)} & 0\\ 0 & 1\end{pmatrix}\otimes\begin{pmatrix}e^{E_\downarrow(N+1)} & 0\\ 0 & 1\end{pmatrix}\otimes\ldots\otimes\begin{pmatrix}e^{E_\downarrow(2N)} & 0\\ 0 & 1\end{pmatrix}.$$

As an abbreviation we have put $-\beta E_\sigma(k)\to E_\sigma(k)$. Since

$$c^\dagger_{k\uparrow}c_{k\uparrow} = I\otimes I\ldots\otimes\overset{k\text{-th place}}{\begin{pmatrix}1 & 0\\ 0 & 0\end{pmatrix}}\otimes I\otimes\ldots\otimes I$$

we have

$$\mathrm{tr}(c^\dagger_{k\uparrow}c_{k\uparrow}W_t)=\frac{\exp(E_\uparrow(k))}{\exp(E_\uparrow(k))+1}.$$

Now we cast the interaction term in a form more convenient for our purpose according to

$$\sum_{k_1k_2k_3k_4} \delta(k_1 - k_2 + k_3 - k_4) c^{\dagger}_{k_1\uparrow} c_{k_2\uparrow} c^{\dagger}_{k_3\downarrow} c_{k_4\downarrow}$$

$$\equiv \sum_{k_1k_3} c^{\dagger}_{k_1\uparrow} c_{k_1\uparrow} c^{\dagger}_{k_3\downarrow} c_{k_3\downarrow} + \sum_{\substack{k_1k_2k_3k_4 \\ k_1 \neq k_2}} \delta(k_1 - k_2 + k_3 - k_4) c^{\dagger}_{k_1\uparrow} c_{k_2\uparrow} c^{\dagger}_{k_3\downarrow} c_{k_4\downarrow}.$$

For the first term on the right-hand side we find

$$c^{\dagger}_{k_1\uparrow} c_{k_1\uparrow} c^{\dagger}_{k_3\downarrow} c_{k_3\downarrow} = I \otimes I \otimes \ldots \otimes \underbrace{\begin{pmatrix} 1 & 0 \\ 0 & 0 \end{pmatrix}}_{k_1\text{-th place}} \otimes I \ldots \otimes I \otimes \underbrace{\begin{pmatrix} 1 & 0 \\ 0 & 0 \end{pmatrix}}_{(k_3+N)\text{-th place}} \otimes I \ldots \otimes I.$$

Thus, it follows that

$$\frac{1}{N} \sum_{k_1k_3} \mathrm{tr}(c^{\dagger}_{k_1\uparrow} c_{k_1\uparrow} c^{\dagger}_{k_3\downarrow} c_{k_3\downarrow} W_t) = \frac{1}{N} \sum_{k_1k_3} \frac{e^{E_{\uparrow}(k_1)}}{1 + e^{E_{\uparrow}(k_1)}} \cdot \frac{e^{E_{\downarrow}(k_3)}}{1 + e^{E_{\downarrow}(k_3)}}.$$

For the second term on the right hand side there is no contribution to the trace since

$$\mathrm{tr}\left(\begin{pmatrix} 0 & -1 \\ 0 & 0 \end{pmatrix} \begin{pmatrix} e^{\lambda} & 0 \\ 0 & 1 \end{pmatrix} \right) = 0$$

where

$$\frac{1}{2}\sigma_+\sigma_z = \begin{pmatrix} 0 & -1 \\ 0 & 0 \end{pmatrix}.$$

The described method can be extended in order to obtain better approximations. To this end we make a unitary transformation given by

$$U := \exp(iS)$$

with $S = S^{\dagger} = S^*$, and followed then by the trace determination. Also, in this case, the trace determination can easily be performed within the framework of usual matrix calculation. ♣

We now calculate the grand canonical partition function Z and grand thermodynamic potential for the Hamilton operator

$$\hat{H} = U \sum_{j=1}^{N} \hat{n}_{j\uparrow} \hat{n}_{j\downarrow}$$

where U is a positive constant (the repulsion of the electrons at the same lattice site j), N is the number of lattice sites and

$$\hat{n}_{j\sigma} := c^{\dagger}_{j\sigma} c_{j\sigma}.$$

For these calculations we need the following theorems

Theorem 1. Let $A_1,\ A_2,\ B_1,\ B_2$ be $n \times n$ matrices. Then

$$\mathrm{tr}\exp(A_1 \otimes I \otimes B_1 \otimes I + I \otimes A_2 \otimes I \otimes B_2) = (\mathrm{tr}\exp(A_1 \otimes B_1))(\mathrm{tr}\exp(A_2 \otimes B_2)).$$

An extension of this formula is

Theorem 2. Let $A_1,\ A_2, \ldots, A_N,\ B_1,\ B_2, \ldots, B_N$ be $n \times n$ matrices. Then

$$\begin{aligned}
&\mathrm{tr}\exp(A_1 \otimes I \otimes \cdots \otimes I \otimes B_1 \otimes I \otimes \cdots \otimes I \\
&\quad + I \otimes A_1 \otimes \cdots \otimes I \otimes I \otimes B_2 \otimes \cdots \otimes I \\
&\quad \vdots \\
&\quad + I \otimes I \otimes \cdots \otimes A_N \otimes I \otimes I \otimes \cdots \otimes B_N) \\
&= (\mathrm{tr}\exp(A_1 \otimes B_1))(\mathrm{tr}\exp A_2 \otimes B_2)) \cdots (\mathrm{tr}\exp(A_N \otimes B_N)).
\end{aligned}$$

For our purpose we need a further extension.

Theorem 3. Let $A_1,\ A_2,\ B_1,\ B_2,$ be $n \times n$ matrices. Assume that

$$[A_1,\ C] = [A_2,\ C] = [B_1,\ C] = [B_2,\ C] = 0.$$

Then

$$\begin{aligned}
&\mathrm{tr}\ \exp(A_1 \otimes I \otimes B_1 \otimes I + I \otimes A_2 \otimes I \otimes B_2 + C \otimes I \otimes I \otimes I \\
&\quad + I \otimes C \otimes I \otimes I + I \otimes I \otimes C \otimes I + I \otimes I \otimes I \otimes C) \\
&= (\mathrm{tr}\ \exp(A_1 \otimes B_1 + C \otimes I + I \otimes C))(\mathrm{tr}\ \exp(A_2 \otimes B_2 + C \otimes I + I \otimes C)).
\end{aligned}$$

An extension is as follows

Theorem 4. Let $A_1, A_2, \ldots, A_N,\ B_1, B_2, \ldots, B_N, C$ be $n \times n$ matrices. Assume that

$$[A_i,\ C] = [B_i,\ C] = 0$$

for $i = 1, 2, \ldots, N$. Then

$$\begin{aligned}
&\operatorname{tr}\exp(A_1 \otimes I \otimes \cdots \otimes I \otimes B_1 \otimes I \otimes \cdots \otimes I \\
&\quad + I \otimes A_2 \otimes \cdots \otimes I \otimes I \otimes B_2 \otimes \cdots \otimes I \\
&\quad \vdots \\
&\quad + I \otimes I \otimes \cdots \otimes A_N \otimes I \otimes I \otimes \cdots \otimes B_N \\
&\quad + C \otimes I \otimes \cdots \otimes I \otimes I \otimes I \otimes \cdots \otimes I \\
&\quad + I \otimes C \otimes \cdots \otimes I \otimes I \otimes I \otimes \cdots \otimes I \\
&\quad \vdots \\
&\quad + I \otimes I \otimes \cdots \otimes I \otimes I \otimes I \otimes \cdots \otimes C) \\
&= \prod_{j=1}^{N} \operatorname{tr}\exp(A_j \otimes B_j + C \otimes I + I \otimes C).
\end{aligned}$$

Before we describe the connection with Fermi systems let us consider a special case of the identity. Assume that

$$A_1 = \cdots = A_N = B_1 = \cdots = B_N.$$

We put

$$A = A_1 = \cdots = B_N.$$

Then the right-hand side of the identity takes the form

$$(\operatorname{tr}\exp(A \otimes A + C \otimes I + I \otimes C))^N.$$

Since

$$[A,\ C] = 0$$

it follows that

$$(\operatorname{tr}\exp(A \otimes A + C \otimes I + I \otimes C))^N \equiv (\operatorname{tr}(\exp(A \otimes A)\exp(C \otimes I)\exp(I \otimes C)))^N.$$

Assume that the matrices A and C can be written as

$$A = \sqrt{a}X, \qquad C = bY$$

where $a \in \mathbf{R}^+$, $b \in \mathbf{R}$ and

$$X^2 = X, \qquad Y^2 = Y.$$

This means that X and Y are idempotent. Then we obtain

$$[\mathrm{tr}\exp(A\otimes A+C\otimes I+I\otimes C)]^N$$

$$=[\mathrm{tr}(I\otimes I+(e^a-1)(X\otimes X))(I\otimes I+(e^b-1)(Y\otimes I))(I\otimes I+(e^b-1)(I\otimes Y))]^N.$$

Using the identity

$$(R\otimes S)(U\otimes V)=(RU)\otimes(SV)$$

it follows that

$$\begin{aligned}&[\mathrm{tr}\exp(A\otimes A+C\otimes I+I\otimes C)]^N\\&\quad=[\mathrm{tr}(I\otimes I+(e^b-1)(Y\otimes I+I\otimes Y)+(e^a-1)(X\otimes X)\\&\quad\quad+(e^b-1)^2(Y\otimes Y)+(e^a-1)(e^b-1)(XY\otimes X+X\otimes XY)\\&\quad\quad+(e^a-1)(e^b-1)^2(XY\otimes XY))]^N.\end{aligned}$$

If we assume that I is the 2×2 unit matrix and

$$X=Y=\begin{pmatrix}1&0\\0&0\end{pmatrix}$$

then

$$(\mathrm{tr}\exp(A\otimes A+C\otimes I+I\otimes C))^N=(1+2e^b+e^{a+2b})^N.$$

Now we have

$$c^\dagger_{j\uparrow}=\underbrace{\sigma_z\otimes\cdots\otimes\sigma_z\otimes\overset{j\text{-th place}}{\left(\frac{1}{2}\sigma_+\right)}\otimes I\otimes\cdots\otimes I}_{2N\text{ times}}$$

$$c^\dagger_{j\downarrow}=\underbrace{\sigma_z\otimes\cdots\otimes\sigma_z\otimes\overset{(j+N)\text{-th place}}{\left(\frac{1}{2}\sigma_+\right)}\otimes I\otimes\cdots\otimes I}_{2N\text{ times}}$$

where $j=1,2,\ldots,2N$, I is the 2×2 unit matrix and

$$\sigma_z:=\begin{pmatrix}1&0\\0&-1\end{pmatrix},\qquad\sigma_+:=\begin{pmatrix}0&2\\0&0\end{pmatrix}.$$

For the Fermi annihilation operators with spin up and down, respectively, we have to replace σ_+ by

$$\sigma_- = \begin{pmatrix} 0 & 0 \\ 2 & 0 \end{pmatrix}.$$

Consider now the Hamilton operator

$$\hat{K} = a\sum_{j=1}^{N} c_{j\uparrow}^\dagger c_{j\uparrow} c_{j\downarrow}^\dagger c_{j\downarrow} + b\sum_{j=1}^{N}(c_{j\uparrow}^\dagger c_{j\uparrow} + c_{j\downarrow}^\dagger c_{j\downarrow})$$

where $a, b \in \mathbf{R}$. Since $\sigma_z^2 = I$ and

$$\left(\frac{1}{2}\sigma_+\right)\left(\frac{1}{2}\sigma_-\right) = \begin{pmatrix} 1 & 0 \\ 0 & 0 \end{pmatrix}$$

we obtain

$$c_{j\uparrow}^\dagger c_{j\uparrow} = I \otimes \cdots \otimes I \otimes \overset{j\text{-th place}}{\begin{pmatrix} 1 & 0 \\ 0 & 0 \end{pmatrix}} \otimes I \otimes \cdots \otimes I$$

$$c_{j\downarrow}^\dagger c_{j\downarrow} = I \otimes \cdots \otimes I \otimes \overset{(j+N)\text{-th place}}{\begin{pmatrix} 1 & 0 \\ 0 & 0 \end{pmatrix}} \otimes I \otimes \cdots \otimes I$$

and

$$c_{j\uparrow}^\dagger c_{j\uparrow} c_{j\downarrow}^\dagger c_{j\downarrow} = I \otimes \cdots \otimes I \otimes \overset{j\text{-th place}}{\begin{pmatrix} 1 & 0 \\ 0 & 0 \end{pmatrix}} \otimes I \otimes \cdots \otimes I \otimes \overset{(j+N)\text{-th place}}{\begin{pmatrix} 1 & 0 \\ 0 & 0 \end{pmatrix}} \otimes I \otimes \cdots \otimes I.$$

Consequently, the problem to calculate $\operatorname{tr}\exp(\hat{K})$ has been solved. Using the equations given above we find that the grand thermodynamic potential per lattice site is given by

$$\frac{\Omega}{N} = -\frac{1}{\beta}\ln(1 + 2\exp(\beta\mu) + \exp(\beta(2\mu - U)))$$

where μ is the chemical potential. The number of electrons per lattice site

$$n_e \equiv \frac{N_e}{N}$$

is defined by

$$n_e := -\frac{1}{N}\frac{\partial\Omega}{\partial\mu}.$$

We find

$$n_e = \frac{2(\exp(\beta\mu) + \exp(\beta(2\mu - U)))}{1 + 2\exp(\beta\mu) + \exp(\beta(2\mu - U))}$$

where $0 \le n_e \le 2$. Because of the Pauli principle we have $\mu \to \infty$ as $n_e \to 2$.

Exercises. (1) Show that

$$[\sigma_z, c^\dagger]_+ = 0, \qquad [\sigma_z, c]_+ = 0.$$

(2) Find the eigenvalues of $\hat{n}_k$ and $\hat{N}_e$.

(3) Let $k = 1, 2, \ldots, N$. Show that $c_k^\dagger$ and c_k can be written as

$$c_k^\dagger = \frac{1}{2}\left(\prod_{j=1}^{k-1} \sigma_{z,j}\right) \sigma_{+,k}, \qquad c_k = \frac{1}{2}\left(\prod_{j=1}^{k-1} \sigma_{z,j}\right) \sigma_{-,k}$$

where

$$\sigma_{z,j} = I \otimes \ldots \otimes I \otimes \sigma_z \otimes I \otimes \ldots \otimes I.$$

(4) Show that

$$\sum_{j=1}^{N} c_j^\dagger c_j = \frac{1}{4}\sum_{j=1}^{N} \sigma_{+,j}\sigma_{-,j}.$$

(5) Let $k = 1, 2, \ldots, N$. Show that

$$c_k^\dagger = \frac{1}{2}(-1)^{k-1}\left(\prod_{j=1}^{k-1} \sigma_{z,j}\right) \sigma_{+,k}, \qquad c_k = \frac{1}{2}(-1)^{k-1}\left(\prod_{j=1}^{k-1} \sigma_{z,j}\right) \sigma_{-,k}$$

is also a faithful representation of the Fermi operators.

(6) Consider the Hamilton operator

$$\hat{H} = W\sum_{j=1}^{N} \hat{n}_j - J\sum_{j=1}^{N}(c_j^\dagger c_{j-1} + c_{j-1}^\dagger c_j) + V\sum_{j=1}^{N} \hat{n}_j \hat{n}_{j-1}$$

with cyclic boundary conditions and $\hat{n}_j := c_j^\dagger c_j$. Here W, J and V are constants. The *Jordan-Wigner transformation* for c_j is defined as

$$c_j = \exp\left(i\pi \sum_{l=1}^{j-1} s_{+,l}s_{-,l}\right) s_{-,j}$$

where

$$s_{-,j} := \frac{1}{2}(\sigma_{x,j} - \sigma_{y,j}) = \frac{1}{2}\sigma_{-,j}, \qquad s_{+,j} := \frac{1}{2}(\sigma_{x,j} + \sigma_{y,j}) = \frac{1}{2}\sigma_{+,j}.$$

Find $c_j^\dagger$ and $\hat{n}_j$. Show that the Hamilton operator takes the form

$$\hat{H} = -\sum_{j=1}^{N}\left(J(s_{+,j}s_{-,j-1} + s_{-,j}s_{+,j-1}) - \frac{V}{4}\sigma_{z,j}\sigma_{z,j-1}\right) + \frac{W+V}{2}\sum_{j=1}^{N}\sigma_{z,j} + \frac{N}{2}\left(W + \frac{V}{2}\right) I.$$

3.4 Dimer Problem

Let M, N be natural numbers. We consider a $2M \times 2N$ square lattice. In how many ways $\mathcal{N}$ can we fill a $2M$ by $2N$ square lattice with non-overlapping dimers ? All sites are occupied and no two dimers overlap. A dimer is a horizontal or vertical bond which occupies two vertices. In other words a dimer is a rigid rot just long enough to cover two neighbouring vertices either horizontally or vertically.

In the following we consider the case of periodic boundary conditions which means that the first column is to be identified with the $2M+1$ th column and the first row with the $2N+1$ th row and $N, M \to \infty$.

There are several methods to solve the dimer problem. We follow the *transfer matrix method* [22], [27]. We have to cast the problem in such a form that we can apply the Kronecker product. We may specify a dimer arrangement by listing the type of dimer at each vertex i.e., up, down, to the left or the right. This introduces the possibility of two dimensional topological problems. To avoid these, we artificially construct composite vertices which can be ordered in a one dimensional space. These composite vertices are simply the rows of the square lattice. The configurations of a given row u form the base vectors of the space in which our computations will be performed with the help of the Kronecker product. It is only necessary to determine whether or not there is a vertical dimer from any given vertex of a row. There are then 2^{2M} base vectors of configurations of a given row. A vertex j with an upward vertical dimer will be specified by the unit vector

$$\begin{pmatrix} 1 \\ 0 \end{pmatrix}$$

or if not occupied by an upward vertical dimer, by the unit vector

$$\begin{pmatrix} 0 \\ 1 \end{pmatrix}.$$

The configuration of u is then given by

$$\begin{aligned} (a_1, \cdots, a_s) \times (b_1, \cdots, b_s) &= (a_1 b_1, \cdots, a_1 b_s, a_2 b_1, \cdots, a_2 b_s, \cdots, a_s b_s) \\ u &= u(1) \otimes u(2) \otimes \cdots \otimes u(2M) \end{aligned}$$

where

$$u(j) = \begin{cases} \begin{pmatrix} 1 \\ 0 \end{pmatrix} & \text{if } j \text{ is occupied by a dimer to its upper neighbour} \\ \begin{pmatrix} 0 \\ 1 \end{pmatrix} & \text{if } j \text{ is not occupied by a dimer to its upper neighbour} \end{cases}$$

and $j = 1, 2, \ldots, 2M$. The configurations of u form an orthonormal basis for the 2^{2M} dimensional vector space which they generate. We impose periodic boundary conditions.

Definition. Suppose that the bottom row, which we call the first, has the configuration u_1. We define the *transfer matrix* $T(w,u)$ between a row u and its upper neighbour w by the condition

$$T(w,u) := \begin{cases} 1 & \text{if } w \text{ and } u \text{ are consistent configurations} \\ 0 & \text{otherwise} \end{cases}$$

By consistent we mean that the designated pair of rows can be continued to a legitimate configuration of the lattice.

The quantity

$$Tu_1 = \sum_{u_2} T(u_2, u_1) u_2$$

is then the sum of all possible configurations of the second row given that the first row is u_1. Analogously

$$T^2 u_1 = T\big(\sum_{u_2} T(u_2,u_1)u_2\big) = \sum_{u_3}\sum_{u_2} T(u_2,u_1)T(u_3,u_2)u_3$$

is the sum of all possible configurations of the third row given that the first is u_1. Iterating this process we see that the sum of all possible configurations of the $2N+1$ th row is given by $T^{2N}u_1$. However according to periodic boundary conditions the configuration of the $2N+1$ th row must be u_1 itself.

The number of possible configurations $\mathcal{N}$ corresponding to u_1 is therefore the coefficient of u_1 in $T^{2N}u_1$ or by orthonormality

$$\mathcal{N}(u_1) = (u_1, T^{2N}u_1).$$

Summing over all u_1 we find that

$$\mathcal{N} = \mathrm{tr}(T^{2N})$$

where tr denotes the trace.

Now we have to evaluate the transfer matrix T. Suppose that we are given a row configuration u and ask for those upper neighbour configurations w which are consistent with u. Obviously, u will interfere with w only by virtue of its vertical dimers which terminate at w. This is why the configuration, as we have defined it, is sufficient to determine consistency. The row w is characterized first of all by its horizontal dimers. Suppose

$$S = \{\,\alpha\,\}$$

is the set of adjacent vertex pairs on which the horizontal dimers of w are placed. A dimer can be placed on w at $(i, i+1)$ only if there are no vertical dimers from u at $(i, i+1)$. This means the partial configuration of u for this pair is given by

$$\begin{pmatrix} 0 \\ 1 \end{pmatrix} \otimes \begin{pmatrix} 0 \\ 1 \end{pmatrix}.$$

Thus the transfer matrix T must yield 0 for each configuration except

$$\begin{pmatrix} 0 \\ 1 \end{pmatrix} \otimes \begin{pmatrix} 0 \\ 1 \end{pmatrix}.$$

Since the $(i, i+1)$ pair on w cannot then be occupied by vertical dimers, T must map

$$\begin{pmatrix} 0 \\ 1 \end{pmatrix} \otimes \begin{pmatrix} 0 \\ 1 \end{pmatrix}$$

into itself. The projection matrix $\bar{H}_\alpha$ which has these properties is given by

$$\begin{array}{rcl} \bar{H}_\alpha & := & I \otimes I \otimes \ldots \otimes I \otimes \begin{pmatrix} 0 & 0 \\ 0 & 1 \end{pmatrix} \otimes \begin{pmatrix} 0 & 0 \\ 0 & 1 \end{pmatrix} \otimes I \otimes \ldots \otimes I \\ & & 1 \quad 2 \qquad \ldots \qquad\quad i \qquad\qquad\quad i+1 \qquad \ldots \qquad 2M \end{array}$$

since

$$\begin{pmatrix} 0 & 0 \\ 0 & 1 \end{pmatrix} \begin{pmatrix} 0 \\ 1 \end{pmatrix} = 1 \begin{pmatrix} 0 \\ 1 \end{pmatrix}$$

and

$$\begin{pmatrix} 0 & 0 \\ 0 & 1 \end{pmatrix} \begin{pmatrix} 1 \\ 0 \end{pmatrix} = 0 \begin{pmatrix} 1 \\ 0 \end{pmatrix}.$$

Here I is the 2×2 unit matrix. Obviously $\bar{H}_\alpha$ is a $2^{2M} \times 2^{2M}$ matrix.

Next we consider a vertex j on w which is not occupied by a horizontal dimer. Its configuration is completely determined by that of the corresponding vertex j on u: a vertical dimer on u means that there cannot be one on w, an absent vertical dimer on u requires a vertical dimer at j on w (since it is not occupied by a horizontal dimer). The transfer matrix must then reverse the configuration of j. The permutation matrix which carries this out takes the form

$$\begin{array}{rcl} P_j & := & I \otimes I \otimes \ldots \otimes I \otimes \begin{pmatrix} 0 & 1 \\ 1 & 0 \end{pmatrix} \otimes I \otimes \ldots \otimes I \\ & & 1 \quad 2 \qquad \ldots \qquad\quad j \qquad\qquad \ldots \qquad 2M \end{array}$$

where I is the 2×2 unit matrix. It follows that

$$T = \sum_S \prod_{\alpha \in S} \bar{H}_\alpha \prod_{j \in \bar{S}} P_j.$$

Using the property (exercise (2))

$$(A \otimes A)(B \otimes I_2)(I_2 \otimes C) = C \otimes C$$

the transfer matrix can be simplified as follows

$$\bar{H}_{i,i+1} = H_{i,i+1} P_i P_{i+1}$$

where

$$H_{i,i+1} = I \otimes \ldots \otimes I \otimes \begin{pmatrix} 0 & 0 \\ 1 & 0 \end{pmatrix} \otimes \begin{pmatrix} 0 & 0 \\ 1 & 0 \end{pmatrix} \otimes I \otimes \ldots \otimes I.$$

Consequently

$$T = \sum_S \prod_{\alpha \in S} H_\alpha \prod_{j=1}^{2M} P_j.$$

Owing to the method of construction, the product over α includes only non-overlapping adjacent-element pairs. However, since

$$\begin{pmatrix} 0 & 0 \\ 1 & 0 \end{pmatrix} \begin{pmatrix} 0 & 0 \\ 1 & 0 \end{pmatrix} = \begin{pmatrix} 0 & 0 \\ 0 & 0 \end{pmatrix}$$

we find that

$$H_\alpha H_\beta = 0 \quad \text{if } \alpha \text{ and } \beta \text{ overlap.}$$

Thus the product over α may be extended to include any set of adjacent pairs whose union is the set S. Consequently

$$T = \prod_{\alpha=(1,2)}^{(2M,1)} (I + H_\alpha) \prod_{j=1}^{2M} P_j = \left(\exp\left(\sum_{i=1}^{2M} H_{i,i+1} \right) \right) \prod_{j=1}^{2M} P_j\,.$$

It will be convenient to consider not T but T^2. Using the results from exercise (3) we find that

$$\begin{aligned} H_{j,j+1} &= s_{-,j} s_{-,j+1} \\ P_j P_{j+1} H_{j,j+1} P_j P_{j+1} &= s_{+,j} s_{+,j+1} \end{aligned}$$

where

$$s_{-,j} := I \otimes \ldots \otimes I \otimes \begin{pmatrix} 0 & 0 \\ 1 & 0 \end{pmatrix} \otimes I \otimes \ldots \otimes I$$

$$s_{+,j} := I \otimes \ldots \otimes I \otimes \begin{pmatrix} 0 & 1 \\ 0 & 0 \end{pmatrix} \otimes I \otimes \ldots \otimes I.$$

Then we find that

$$T^2 = \exp \sum_{j=1}^{2M} (s_{-,j}s_{-,j+1}) \exp \sum_{j=1}^{2M} (s_{+,j}s_{+,j+1}) \, .$$

Let $\lambda_{\max}$ be the maximum eigenvalue of T^2. In the asymptotic limit $N \to \infty$,

$$\mathcal{N} = \mathrm{tr}(T^2)^N \simeq (\lambda_{\max}(T^2))^N$$

so that we only have to find the maximum eigenvalue $\lambda_{\max}$ of T^2.

To find the maximum eigenvalue of T^2 we apply the *Paulion to Fermion Transformations.* This is done by appending a sign which depends multiplicatively on the states of the preceding vertices. We define

$$c_k := \prod_{j=1}^{k-1}(-\sigma_{z,j})s_{-,k}, \qquad c_k^* := \prod_{j=1}^{k-1}(-\sigma_{z,j})s_{+,k}$$

where

$$\sigma_{z,k} = I \otimes \ldots \otimes I \otimes \begin{pmatrix} 1 & 0 \\ 0 & -1 \end{pmatrix} \otimes I \otimes \ldots \otimes I \, .$$

Here we use the notation c_k^* instead of $c_k^\dagger$. We find that

$$s_{-,j}s_{-,j+1} = -c_j c_{j+1}, \qquad j = 1, \ldots, 2M-1$$

$$s_{+,j}s_{+,j+1} = c_j^* c_{j+1}^*, \qquad j = 1, \ldots, 2M-1.$$

On the other hand for $j = 2M$ it is necessary to introduce the Hermitian matrix

$$A := \sum_{j=1}^{2M} c_j^* c_j = \sum_{j=1}^{2M} s_{+,j}s_{-,j}.$$

Using the result from exercise (5) we find

$$T^2 = \exp\left(-\sum_{j=1}^{2M-1} c_j c_{j+1} + (-1)^A c_{2M}c_1 \right) \exp\left(\sum_{j=1}^{2M-1} c_j^* c_{j+1}^* - (-1)^A c_{2M}^* c_1^* \right).$$

where $(-1)^A \equiv \exp(i\pi A)$. Now the eigenvector belonging to the maximum eigenvalue of T^2 is also an eigenvector of A. The corresponding eigenvalue of the matrix A is odd. Thus the matrix T^2 takes the form

$$T^2 = \exp\left(-\sum_{j=1}^{2M} c_j c_{j+1} \right) \exp\left(\sum_{j=1}^{2M} c_j^* c_{j+1}^* \right).$$

To diagonalize T^2, we introduce the *discrete Fourier transformation*

$$c_j = \frac{1}{(2Mi)^{1/2}} \sum_{k=k_0}^{2Mk_0} \exp(ijk) S_k, \qquad c_j^* = \frac{1}{(-2Mi)^{1/2}} \sum_{k=k_0}^{2Mk_0} \exp(-ijk) S_k^*$$

where

$$k_0 = \frac{2\pi}{2M}.$$

It follows that

$$S_k S_\ell^* + S_\ell^* S_k = \frac{1}{2M} \sum_{j,j'} \exp(-ijk + ij'\ell)(c_j c_{j'}^* + c_{j'}^* c_j) = \frac{1}{2M} \sum_{j,j'} \exp(-ijk + ij'\ell)\delta_{j,j'} = \delta_{k\ell}.$$

All other anticommutators vanish. The exponents of T^2 again become diagonal in the S_k's, resulting in

$$T^2 = \prod_{k=k_0}^{Mk_0} \Lambda_k$$

where

$$\Lambda_k = \exp\left(2S_k S_{-k} \sin k\right) \exp\left(2S_{-k}^* S_k^* \sin k\right).$$

To find the maximum eigenvalue of T^2, we can restrict our attention to a special subspace which is invariant under T^2. It is defined by first choosing any vector $\mathbf{x}_0$ which satisfies

$$S_k \mathbf{x}_0 = \mathbf{0}$$

for all k, and then constructing the unit vectors

$$\mathbf{e}(\delta_1, \ldots, \delta_m) = \prod_{k=k_0}^{Mk_0} (S_{-k}^* S_k^*)^{\delta_k} \mathbf{x}_0$$

$$\delta_k = 0 \;\text{ or }\; 1.$$

We find that

$$S_{-k}^* S_k^* = I \otimes \ldots \otimes I \otimes \begin{pmatrix} 0 & 0 \\ 1 & 0 \end{pmatrix} \otimes I \otimes \ldots \otimes I$$

$$S_k S_{-k} = I \otimes \ldots \otimes I \otimes \begin{pmatrix} 0 & 1 \\ 0 & 0 \end{pmatrix} \otimes I \otimes \ldots \otimes I$$

where I is the 2×2 unit matrix. Hence we find

$$\begin{aligned}
\Lambda_k &= I \otimes \ldots \otimes I \otimes \exp\left(2\begin{pmatrix} 0 & 0 \\ 1 & 0 \end{pmatrix}\sin k\right)\exp\left(2\begin{pmatrix} 0 & 1 \\ 0 & 0 \end{pmatrix}\sin k\right) \otimes I \otimes \ldots \otimes I \\
&= I \otimes \ldots \otimes I \otimes \left[I + 2\begin{pmatrix} 0 & 0 \\ 1 & 0 \end{pmatrix}\sin k\right]\left[I + 2\begin{pmatrix} 0 & 1 \\ 0 & 0 \end{pmatrix}\sin k\right] \otimes I \otimes \ldots \otimes I \\
&= I \otimes \ldots I \otimes \begin{pmatrix} 1 & 2\sin k \\ 2\sin k & 1 + 4\sin^2 k \end{pmatrix} \otimes I \otimes \ldots \otimes I.
\end{aligned}$$

The maximum eigenvalue is found to be

$$\left(\sin k + (1 + \sin^2 k)^{1/2}\right)^2.$$

Thus the number of configurations is given by

$$\begin{aligned}
\mathcal{N}(2M, 2N) &\simeq \left[\lambda_{\max}(T^2)\right]^N \\
&= \prod_{k=k_0}^{Mk_0} \left[\sin k + (1 + \sin^2 k)^{1/2}\right]^{2N} \\
&= \exp 2N \sum_{k=k_0}^{Mk_0} \ln(\sin k + (1 + \sin^2 k)^{1/2}).
\end{aligned}$$

As $M \to \infty$ we obtain

$$\mathcal{N} = \exp\left(\frac{2NM}{\pi}\int_0^\pi \ln(\sin k + (1 + \sin^2 k)^{1/2})dk\right).$$

The integration yields

$$\mathcal{N}(2M, 2N) \simeq \exp\left(\frac{G}{\pi}(2M)(2N)\right)$$

where

$$G = 0.915965594\ldots \quad \text{Catalan's constant.}$$

Exercises. (1) Let $N = M = 1$. Find $\mathcal{N}$. Let $N = M = 2$. Find $\mathcal{N}$.

(2) Let

$$A = \begin{pmatrix} 0 & 0 \\ 1 & 0 \end{pmatrix}, \qquad B = \begin{pmatrix} 0 & 1 \\ 1 & 0 \end{pmatrix}, \qquad C = \begin{pmatrix} 0 & 0 \\ 0 & 1 \end{pmatrix}.$$

Show that

$$(A \otimes A)(B \otimes I_2)(I_2 \otimes C) = C \otimes C.$$

(3) Show that

$$P_j P_j = I_{2^{2M} \times 2^{2M}}$$

and

$$P_j P_{j+1} H_{j,j+1} P_j P_{j+1} = I \otimes \ldots \otimes I \otimes \begin{pmatrix} 0 & 1 \\ 0 & 0 \end{pmatrix} \otimes \begin{pmatrix} 0 & 1 \\ 0 & 0 \end{pmatrix} \otimes I \otimes \ldots \otimes I.$$

(4) Show that T^2 is a Hermitian matrix.

(5) Show that

$$c_j^* c_k^* + c_k^* c_j^* = 0, \qquad c_j c_k + c_k c_j = 0, \qquad c_j^* c_k + c_k c_j^* = \delta_{jk} I$$

where $j, k = 1, 2, \ldots, 2M$.

(6) Show that the eigenvalues of the Hermitian matrix A are integers, that A commutes with T^2, that the matrix $(-1)^A$ commutes with any monomial of even degree in the c_j's and c_j^*'s and that

$$s_{-,2M} s_{-,1} = (-1)^A c_{2M} c_1, \qquad s_{+,2M} s_{+,1} = -(-1)^A c_{2M}^* c_1^*.$$

(7) Show that $S_k S_{-k}$ and $S_{-k}^* S_k^*$ do commute with $S_\ell S_{-\ell}$ and $S_{-\ell}^* S_\ell^*$ when $\ell \neq \pm k$.

(8) Show that the number of ways to cover a chessboard with dominoes, each domino filling two squares is given by 12 988 816.

3.5 Two-Dimensional Ising Model

In section 3.2 we have already studied the one-dimensional Ising model. Here we discuss the two-dimensional Ising model. The model considered is a plane square lattice having N points, at each of which is a spin with its axis perpendicular to the lattice plane. The spin can have two opposite orientations, so that the total number of possible configurations of the spins in the lattice is 2^N. To each lattice point (with integral coordinates k, l) we assign a variable σ_{kl} which takes two values ± 1, corresponding to the two possible orientations of the spin. If we take into account only the interaction between adjoining spins, the energy of the configuration is given by

$$E(\sigma) := -J \sum_{k,l=1}^{n} (\sigma_{kl}\sigma_{k,l+1} + \sigma_{kl}\sigma_{k+1,l})$$

where n is the number of points in a lattice line, the lattice being considered as a large square, and $N = n^2$.

In our approach we follow Onsager [26] (see also Kaufman [18], Huang [16]) since the Kronecker product is used. Consider a square lattice of

$$N = n^2$$

spins consisting of n rows and n columns. Let us imagine the lattice to be enlarged by one row and one column with the requirement that the configuration of the $(n+1)$th row and column be identical with that of the first row and column respectively. This boundary condition endows the lattice with the topology of a torus. Let μ_α $(\alpha = 1, \ldots, n)$ denote the collection of all the spin coordinates of the α-th row

$$\mu \equiv \{\sigma_1, \sigma_2, \cdots, \sigma_n\} \quad \alpha\text{-th row}.$$

The toroidal (periodic) boundary condition implies the definition

$$\mu_{n+1} \equiv \mu_1.$$

A configuration of the entire lattice is then specified by

$$\{\mu_1, \cdots, \mu_n\}.$$

There are 2^N configurations. By assumption, the α-th row interacts only with the $(\alpha - 1)$-th and the $(\alpha + 1)$-th row. Let $E(\mu_\alpha, \mu_{\alpha+1})$ be the interaction energy between the α-th and the $(\alpha+1)$-th row. Let $E(\mu_\alpha)$ be the interaction energy of the spins within the α-th row. We can write

$$E(\mu, \mu') = -J \sum_{k=1}^{n} \sigma_k \sigma'_k, \qquad E(\mu) = -J \sum_{k=1}^{n} \sigma_k \sigma_{k+1}$$

where μ and μ' respectively denote the collection of spin coordinates in two neighbouring rows

$$\mu \equiv \{\sigma_1, \cdots, \sigma_n\}, \qquad \mu' \equiv \{\sigma'_1, \cdots, \sigma'_n\}.$$

The toroidal boundary condition implies that in each row

$$\sigma_{n+1} \equiv \sigma_1.$$

The total energy of the lattice for the configuration $\{\mu_1, \ldots, \mu_n\}$ is then given by

$$E_I\{\mu_1, \cdots, \mu_n\} = \sum_{\alpha=1}^{n} [E(\mu_\alpha, \mu_{\alpha+1}) + E(\mu_\alpha)].$$

The *partition function* $Z(\beta)$ is defined as

$$Z(\beta) := \sum_{\mu_1} \cdots \sum_{\mu_n} \exp\left(-\beta \sum_{\alpha=1}^{n} (E(\mu_\alpha, \mu_{\alpha+1}) + E(\mu_\alpha))\right).$$

Let a $2^n \times 2^n$ matrix P be so defined that its matrix elements are

$$\langle \mu | P | \mu' \rangle = \exp\left(-\beta(E(\mu, \mu') + E(\mu))\right).$$

Using this definition we find

$$Z(\beta) = \sum_{\mu_1} \cdots \sum_{\mu_n} \langle \mu_1 | P | \mu_2 \rangle \langle \mu_2 | P | \mu_3 \rangle \cdots \langle \mu_n | P | \mu_1 \rangle = \sum_{\mu_1} \langle \mu_1 | P^n | \mu_1 \rangle = \mathrm{tr} P^n$$

where tr denotes the trace. Since the trace of a matrix is independent of the representation of the matrix, the trace may be evaluated by bringing the symmetric matrix P into its diagonal form, i.e.

$$P = \mathrm{diag}(\lambda_1, \lambda_2, \ldots, \lambda_{2^n})$$

where

$$\lambda_1, \ \lambda_2, \ \cdots, \ \lambda_{2^n}$$

are the 2^n real eigenvalues of P. Notice that the eigenvalues depend on β. The matrix P^n is then also diagonal, with the diagonal matrix elements

$$(\lambda_1)^n, \quad (\lambda_2)^n, \quad \cdots, \quad (\lambda_{2^n})^n.$$

It follows that

$$Z(\beta) = \sum_{\alpha=1}^{2^n} (\lambda_\alpha)^n.$$

We expect that the eigenvalues of the matrix P are in general of the order of $\exp(n)$ when n is large, since $E(\mu, \mu')$ and $E(\mu)$ are of the order of n. If $\lambda_{\max}$ is the largest eigenvalue of P, it follows that

$$\lim_{n \to \infty} \frac{1}{n} \ln \lambda_{\max} = \text{finite number}.$$

If all the eigenvalues λ_α are positive, we obtain

$$(\lambda_{\max})^n \leq Z(\beta) \leq 2^n (\lambda_{\max})^n$$

or

$$\frac{1}{n} \ln \lambda_{\max} \leq \frac{1}{n^2} \ln Z(\beta) \leq \frac{1}{n} \ln \lambda_{\max} + \frac{1}{n} \ln 2.$$

Therefore

$$\lim_{N \to \infty} \frac{1}{N} \ln Z(\beta) = \lim_{n \to \infty} \frac{1}{n} \ln \lambda_{\max}$$

where $N = n^2$. We find that this is true and that all the eigenvalues λ_α are positive. Thus it is sufficient to find the largest eigenvalue of the matrix P.

We obtain the matrix elements of P in the form

$$\langle \sigma_1, \cdots, \sigma_n | P | \sigma'_1, \cdots, \sigma'_n \rangle = \prod_{k=1}^{n} \exp(\beta J \sigma_k \sigma_{k+1}) \exp(\beta J \sigma_k \sigma'_k).$$

We define two $2^n \times 2^n$ matrices V'_1 and V_2 whose matrix elements are respectively given by

$$\langle \sigma_1, \cdots, \sigma_n | V'_1 | \sigma'_1, \cdots, \sigma'_n \rangle = \prod_{k=1}^{n} \exp(\beta J \sigma_k \sigma'_k)$$

$$\langle \sigma_1, \cdots, \sigma_n | V_2 | \sigma'_1, \cdots, \sigma'_n \rangle = \delta_{\sigma_1 \sigma'_1} \cdots \delta_{\sigma_n \sigma'_n} \prod_{k=1}^{n} \exp(\beta J \sigma_k \sigma_{k+1})$$

where $\delta_{\sigma,\sigma'}$ is the Kronecker symbol. Thus V_2 is a diagonal matrix in the present representation. Then it follows that

$$P = V_2 V'_1.$$

We now introduce some special matrices in terms of which V'_1, and V_2 may be expressed. Let σ_x, σ_y and σ_z be the three Pauli spin matrices. The $2^n \times 2^n$ matrices $\sigma_{x,\alpha}$, $\sigma_{y,\alpha}$, $\sigma_{z,\alpha}$ are defined as

$$\begin{aligned}
\sigma_{x,\alpha} &:= I \otimes I \otimes \cdots \otimes \sigma_x \otimes \cdots \otimes I, \qquad n \text{ factors} \\
\sigma_{y,\alpha} &:= I \otimes I \otimes \cdots \otimes \sigma_y \otimes \cdots \otimes I, \qquad n \text{ factors} \\
\sigma_{z,\alpha} &:= I \otimes I \otimes \cdots \otimes \underset{\substack{\uparrow \\ \alpha\text{-th factor}}}{\sigma_z} \otimes \cdots \otimes I, \qquad n \text{ factors}
\end{aligned}$$

where $\alpha = 1, 2, \ldots, n$ and I is the 2×2 unit matrix.

The matrix V'_1 is a Kronecker product of n 2×2 identical matrices

$$V_1' = A \otimes A \otimes \cdots \otimes A$$

where

$$\langle \sigma | A | \sigma' \rangle = \exp(\beta J \sigma \sigma').$$

Since $\sigma, \sigma' \in \{1, -1\}$ we have

$$A = \begin{pmatrix} \exp(\beta J) & \exp(-\beta J) \\ \exp(-\beta J) & \exp(\beta J) \end{pmatrix} \equiv \exp(\beta J) I + \exp(-\beta J) \sigma_x$$

where I is the 2×2 unit matrix. Obviously, A can be written as

$$A = \sqrt{2 \sinh(2\beta J)} \exp(\theta \sigma_x)$$

where

$$\tanh \theta \equiv \exp(-2\beta J).$$

It follows that

$$V_1' = (2 \sinh(2\beta J))^{n/2} \exp(\theta \sigma_x) \otimes \exp(\theta \sigma_x) \otimes \cdots \otimes \exp(\theta \sigma_x).$$

Applying the identity

$$\exp(\theta \sigma_x) \otimes \exp(\theta \sigma_x) \otimes \cdots \otimes \exp(\theta \sigma_x) \equiv \exp(\theta \sigma_{x,1}) \exp(\theta \sigma_{x,2}) \cdots \exp(\theta \sigma_{x,n})$$

we obtain

$$V_1' = (2 \sinh(2\beta J))^{n/2} V_1, \qquad V_1 = \prod_{\alpha=1}^{n} \exp(\theta \sigma_{x,\alpha}).$$

For the matrix V_2 we find

$$V_2 = \prod_{\alpha=1}^{n} \exp(\beta J \sigma_{z,\alpha} \sigma_{z,\alpha+1})$$

where

$$\sigma_{z,n+1} = \sigma_{z,1}.$$

Therefore

$$P = (2 \sinh(2\beta J))^{n/2} V_2 V_1.$$

This completes the formulation of the two-dimensional Ising model.

We introduce a class of matrices which are relevant to the solution of the two-dimensional Ising model.

Let $2n$ matrices Γ_μ, $\mu = 1, \ldots, 2n$, be defined as a set of matrices satisfying the following anticommutation rule

$$\Gamma_\mu \Gamma_\nu + \Gamma_\nu \Gamma_\mu = 2\delta_{\mu\nu} I$$

where $\nu = 1, \ldots, 2n$. A representation of the matrices $\{\Gamma_\mu\}$ by $2^n \times 2^n$ matrices is given by

$$\begin{array}{ll} \Gamma_1 = \sigma_{z,1} & \Gamma_2 = \sigma_{y,1} \\ \Gamma_3 = \sigma_{x,1}\sigma_{z,2} & \Gamma_4 = \sigma_{x,1}\sigma_{y,2} \\ \Gamma_5 = \sigma_{x,1}\sigma_{x,2}\sigma_{z,3} & \Gamma_6 = \sigma_{x,1}\sigma_{x,2}\sigma_{y,3} \\ \vdots & \vdots \end{array}$$

Therefore

$$\begin{aligned} \Gamma_{2\alpha-1} &= \sigma_{x,1}\sigma_{x,2}\cdots\sigma_{x,\alpha-1}\sigma_{z,\alpha}, \qquad \alpha = 1, \ldots, n \\ \Gamma_{2\alpha} &= \sigma_{x,1}\sigma_{x,2}\ldots\sigma_{x,\alpha-1}\sigma_{y,\alpha}, \qquad \alpha = 1, \ldots, n \end{aligned}$$

Another representation is obtained by interchanging the roles of $\sigma_{x,\alpha}$ and $\sigma_{y,\alpha}$ ($\alpha = 1, \ldots, n$). Another representation results by an arbitrary permutation of the numbering of $\Gamma_1, \ldots, \Gamma_{2n}$. The $2^n \times 2^n$ matrices V_1 and V_2 are matrices that transform one set of $\{\Gamma_\mu\}$ into another equivalent set. Let a definite set $\{\Gamma_\mu\}$ be given and let ω be the $2n \times 2n$ matrix describing a linear orthogonal transformation among the members of $\{\Gamma_\mu\}$, i.e.

$$\Gamma'_\mu = \sum_{\nu=1}^{2n} \omega_{\mu\nu} \Gamma_\nu$$

where $\omega_{\mu\nu}$ are complex numbers satisfying

$$\sum_{\mu=1}^{2n} \omega_{\mu\nu}\omega_{\mu\lambda} = \delta_{\nu\lambda}.$$

This relation can be written in matrix form as

$$\omega^T \omega = I$$

where ω^T is the transpose of ω and I is the $2n \times 2n$ unit matrix. If Γ_μ is regarded as a component of a vector in a $2n$-dimensional space, then ω induces a rotation in that space

$$\begin{pmatrix} \Gamma'_1 \\ \Gamma'_2 \\ \vdots \\ \Gamma'_{2n} \end{pmatrix} = \begin{pmatrix} \omega_{11} & \omega_{12} & \cdots & \omega_{1,2n} \\ \omega_{21} & \omega_{22} & \cdots & \omega_{2,2n} \\ \vdots & & & \\ \omega_{2n,1} & \omega_{2n,2} & \cdots & \omega_{2n,2n} \end{pmatrix} \begin{pmatrix} \Gamma_1 \\ \Gamma_2 \\ \vdots \\ \Gamma_{2n} \end{pmatrix}.$$

Therefore

$$\Gamma'_\mu = S(\omega)\Gamma_\mu S^{-1}(\omega)$$

where $S(\omega)$ is a nonsingular $2^n \times 2^n$ matrix. The existence of $S(\omega)$ will be demonstrated by explicit construction. Thus there is a mapping

$$\omega \leftrightarrow S(\omega)$$

which establishes $S(\omega)$ as a $2^n \times 2^n$ matrix representation of a rotation in a $2n$-dimensional space. It follows that

$$S(\omega)\Gamma_\mu S^{-1}(\omega) = \sum_{\nu=1}^{2n} \omega_{\mu\nu}\Gamma_\nu .$$

We call ω a *rotation* and $S(\omega)$ the spin representative of the rotation ω. If ω_1 and ω_2 are two rotations then $\omega_1\omega_2$ is also a rotation. Furthermore

$$S(\omega_1\omega_2) = S(\omega_1)S(\omega_2).$$

We now study some special rotations ω and their corresponding $S(\omega)$. Consider a rotation in a two-dimensional plane of the $2n$-dimensional space. A rotation in the plane $\mu\nu$ through the angle θ is defined by the transformation

$$\begin{aligned}
\Gamma'_\lambda &= \Gamma_\lambda & &(\lambda \neq \mu, \lambda \neq \nu) \\
\Gamma'_\mu &= \Gamma_\mu \cos\theta - \Gamma_\nu \sin\theta & &(\mu \neq \nu) \\
\Gamma'_\nu &= \Gamma_\mu \sin\theta + \Gamma_\nu \cos\theta & &(\mu \neq \nu)
\end{aligned}$$

where θ is a complex number. The rotation matrix, denoted by $\omega(\mu\nu|\theta)$, is defined by

$$\omega(\mu\nu|\theta) := \begin{pmatrix} & \vdots & & \vdots & \\ \cdots & \cos\theta & \cdots & \sin\theta & \cdots \\ & \vdots & & \vdots & \\ \cdots & -\sin\theta & \cdots & \cos\theta & \cdots \\ & \vdots & & \vdots & \end{pmatrix}$$

where the matrix elements $\cos\theta$ are in the μ-th column and μ-th row and ν-th column and ν-th row, respectively. The matrix element $\sin\theta$ is in the ν-th column and μ-th row and the matrix element $-\sin\theta$ is in the μ-th column and ν-th row. The matrix elements not displayed are unity along the diagonal and zero everywhere else. Thus the matrix $\omega(\mu,\nu|\theta)$ is a *Givens matrix*. The matrix $\omega(\mu\nu|\theta)$ is called the plane rotation in the plane $\mu\nu$.

The properties of ω and $S(\omega)$ that are relevant to the solution of the two-dimensional Ising model are summarized in the following lemmas.

Lemma 1. If

$$\omega(\mu\nu|\theta) \leftrightarrow S_{\mu\nu}(\theta)$$

then

$$S_{\mu\nu}(\theta) = \exp\left(-\frac{1}{2}\theta\Gamma_\mu\Gamma_\nu\right).$$

Proof. Since $\Gamma_\mu\Gamma_\nu = -\Gamma_\nu\Gamma_\mu$ for $\mu \neq \nu$ we have

$$\Gamma_\mu\Gamma_\nu\Gamma_\mu\Gamma_\nu = -I.$$

Therefore we have

$$\exp\left(-\frac{1}{2}\theta\Gamma_\mu\Gamma_\nu\right) = I\cos\frac{\theta}{2} - \Gamma_\mu\Gamma_\nu\sin\frac{\theta}{2}.$$

Since

$$(\Gamma_\mu\Gamma_\nu)(\Gamma_\nu\Gamma_\mu) = (\Gamma_\nu\Gamma_\mu)(\Gamma_\mu\Gamma_\nu) = I$$

we find

$$\exp\left(\frac{1}{2}\theta\Gamma_\mu\Gamma_\nu\right)\exp\left(-\frac{1}{2}\theta\Gamma_\mu\Gamma_\nu\right) = I.$$

Hence

$$S_{\mu\nu}^{-1}(\theta) = \exp\left(\frac{1}{2}\theta\Gamma_\mu\Gamma_\nu\right).$$

A straightforward calculation shows that

$$\begin{aligned}
S_{\mu\nu}(\theta)\Gamma_\lambda S_{\mu\nu}^{-1}(\theta) &= \Gamma_\lambda, \qquad \lambda \neq \mu, \lambda \neq \nu \\
S_{\mu\nu}(\theta)\Gamma_\mu S_{\mu\nu}^{-1}(\theta) &= \Gamma_\mu\cos\theta + \Gamma_\nu\sin\theta \\
S_{\mu\nu}(\theta)\Gamma_\nu S_{\mu\nu}^{-1}(\theta) &= \Gamma_\mu\sin\theta - \Gamma_\nu\cos\theta\,.
\end{aligned}$$

♠

Lemma 2. The eigenvalues of the $2^n \times 2^n$ matrix $\omega(\mu\nu|\theta)$ are 1 ($2n-2$-fold degenerate), and $\exp(\pm i\theta)$ (nondegenerate). The eigenvalues of $\mathrm{S}_{\mu\nu}(\theta)$ are $\exp(\pm i\theta/2)$ (each 2^{n-1}-fold degenerate).

Proof. The first part is trivial. The second part can be proved by choosing a special representation for $\Gamma_\mu\Gamma_\nu$, since the eigenvalues of $S_{\mu\nu}(\theta)$ are independent of the representation. As a representation for Γ_μ and Γ_ν we use the representation given above with σ_x and σ_z interchanged. We choose

$$\Gamma_\mu = \sigma_{z,1}\sigma_{x,2}, \qquad \Gamma_\nu = \sigma_{z,1}\sigma_{y,2}.$$

Then

$$\Gamma_\mu \Gamma_\nu = \sigma_{x,2}\sigma_{y,2} = i\sigma_{z,2} = I_2 \otimes \begin{pmatrix} i & 0 \\ 0 & -i \end{pmatrix} \otimes I_2 \otimes \cdots \otimes I_2$$

since $\sigma_{z,1}\sigma_{z,1} = I$. Therefore

$$S_{\mu\nu}(\theta) = I\cos\frac{\theta}{2} - \Gamma_\mu\Gamma_\nu \sin\frac{\theta}{2} = I_2 \otimes \begin{pmatrix} e^{-i\theta/2} & 0 \\ 0 & e^{i\theta/2} \end{pmatrix} \otimes I_2 \otimes \cdots \otimes I_2.$$

Consequently, $S_{\mu\nu}(\theta)$ is diagonal. The diagonal elements are either $\exp(i\theta/2)$ or $\exp(-i\theta/2)$, each appearing 2^{n-1} times each. ♠

Lemma 3. Let ω be a product of n commuting plane rotations

$$\omega = \omega(\alpha\beta|\theta_1)\omega(\gamma\delta|\theta_2)\cdots\omega(\mu\nu|\theta_n)$$

where $\{\alpha, \beta, \ldots, \mu, \nu\}$ is a permutation of the set of integers $\{1, 2, \ldots, 2n-1, 2n\}$, and $\theta_1, \ldots, \theta_n$ are complex numbers. Then
(a) $\omega \leftrightarrow S(\omega)$, with

$$S(\omega) = \exp\left(-\frac{1}{2}\theta_1\Gamma_\alpha\Gamma_\beta\right)\exp\left(-\frac{1}{2}\theta_2\Gamma_\gamma\Gamma_\delta\right)\cdots\exp\left(-\frac{1}{2}\theta_n\Gamma_\mu\Gamma_\nu\right).$$

(b) The $2n$ eigenvalues of ω are given by

$$\exp(\pm i\theta_1), \quad \exp(\pm i\theta_2), \ldots, \exp(\pm i\theta_n).$$

(c) The 2^n eigenvalues of $S(\omega)$ are the values

$$\exp\left(\frac{1}{2}i(\pm\theta_1 \pm \theta_2 \pm \cdots \pm \theta_n)\right)$$

with the signs $\pm$ are chosen independently.

Proof. This lemma is a consequence of lemmas 1 and 2 and the fact that

$$[\Gamma_\mu\Gamma_\nu, \Gamma_\alpha\Gamma_\beta] = 0. \qquad ♠$$

By this lemma, the eigenvalues of $S(\omega)$ can be obtained from those of ω. With the help of these lemmas we can express V_2V_1 in terms of $S(\omega)$.

Now we can find the solution. The partition function is given by

$$\lim_{N\to\infty}\frac{1}{N}\ln Z(\beta) = \frac{1}{2}\ln[2\sinh(2\beta J)] + \lim_{n\to\infty}\frac{1}{n}\ln\Lambda$$

where

$$\Lambda = \text{largest eigenvalue of } V$$

and

$$V = V_1 V_2.$$

These formulas are valid if all eigenvalues of V are positive and if

$$\lim_{n\to\infty} \frac{1}{n} \ln \Lambda$$

exists. Thus we have to diagonalize the matrix V. Using the representation given above, we find that

$$\Gamma_{2\alpha}\Gamma_{2\alpha-1} = \sigma_{y,\alpha}\sigma_{z,\alpha} = i\sigma_{x,\alpha}, \qquad \alpha = 1, \ldots, n$$

and

$$V_1 = \prod_{\alpha=1}^{n} \exp(\theta\sigma_{x,\alpha}) = \prod_{\alpha=1}^{n} \exp(-i\theta\Gamma_{2\alpha}\Gamma_{2\alpha-1}).$$

Thus V_1 is a spin representative of a product of commuting plane rotations. Furthermore

$$\Gamma_{2\alpha+1}\Gamma_{2\alpha} = \sigma_{x,\alpha}\sigma_{z,\alpha+1}\sigma_{y,\alpha} = i\sigma_{z,\alpha}\sigma_{z,\alpha+1} \qquad \alpha = 1, \ldots, n-1$$

$$\Gamma_1\Gamma_{2n} = \sigma_{z,1}(\sigma_{x,1}\cdots\sigma_{x,n-1})\sigma_{y,n} = -i\sigma_{z,1}\sigma_{z,n}(\sigma_{x,1}\cdots\sigma_{x,n}).$$

We have

$$V_2 = \left(\prod_{\alpha=1}^{n-1} \exp(\beta J\sigma_{z,\alpha}\sigma_{z,\alpha+1})\right) \exp(\beta J\sigma_{z,n}\sigma_{z,1}).$$

Therefore we can write

$$V_2 = \exp(i\beta JU\Gamma_1\Gamma_{2n}) \prod_{\alpha=1}^{n-1} \exp(-i\beta J\Gamma_{2\alpha+1}\Gamma_{2\alpha})$$

where the $2^n \times 2^n$ matrix U is defined by

$$U := \sigma_{x,1}\sigma_{x,2}\cdots\sigma_{x,n}.$$

Owing to the first factor in $\exp(i\beta JU\Gamma_1\Gamma_2)$, the matrix V_2 is not a spin representative of a product of commuting plane rotations. This factor owes its existence to the toroidal boundary condition imposed on the problem (i.e., the condition that $\sigma_{n+1} \equiv \sigma_1$ in every row of the lattice). Thus for the matrix $V = V_2V_1$ we obtain

$$V = V_2V_1 = \exp(i\phi U\Gamma_1\Gamma_{2n}) \left(\prod_{\alpha=1}^{n-1} \exp(-i\phi\Gamma_{2\alpha+1}\Gamma_{2\alpha})\right) \left(\prod_{\lambda=1}^{n} \exp(-i\phi\Gamma_{2\lambda}\Gamma_{2\lambda-1})\right)$$

where

$$\phi = \beta J, \qquad J > 0, \qquad \theta = \tanh^{-1}\exp(-2\phi).$$

A straightforward calculation shows that

$$\exp(i\phi\Gamma_1\Gamma_{2n}U) == \frac{1}{2}(I+U)\exp(i\phi\Gamma_1\Gamma_{2n}) + \frac{1}{2}(I-U)\exp(-i\phi\Gamma_1\Gamma_{2n}).$$

Substituting this result into V we obtain

$$V = \frac{1}{2}(I+U)V^+ + \frac{1}{2}(I-U)V^-$$

where

$$V^\pm = \exp(\pm i\phi\Gamma_1\Gamma_{2n})\left(\prod_{\alpha=1}^{n-1}\exp(-i\phi\Gamma_{2\alpha+1}\Gamma_{2\alpha})\right)\left(\prod_{\lambda=1}^{n}\exp(-i\theta\Gamma_{2\lambda}\Gamma_{2\lambda-1})\right).$$

Thus both matrices V^+ and V^- are spin representatives of rotations.

Now we look for a representation in which U is diagonal. The three matrices U, V^+ and V^- commute with one another. Consequently, the three matrices U, V^+, V^- can be simultaneously diagonalized. We first transform V into the representation in which U is diagonal (but in which $V^\pm$ are not necessarily diagonal)

$$RVR^{-1} \equiv \widetilde{V} = \frac{1}{2}(I+\widetilde{U})\widetilde{V}^+ + \frac{1}{2}(I-\widetilde{U})\widetilde{V}^-$$

$$\widetilde{U} \equiv RUR^{-1}, \qquad \widetilde{V}^\pm \equiv RV^\pm R^{-1}.$$

Since $U^2 = I$ the eigenvalues of U are either $+1$ or -1. We find that U can also be written in the form

$$U = \sigma_x \otimes \sigma_x \otimes \cdots \otimes \sigma_x.$$

Therefore a diagonal form of U is

$$\sigma_z \otimes \sigma_z \otimes \cdots \otimes \sigma_z$$

and the eigenvalues are 1 2^{n-1} times and -1 2^{n-1} times. Other diagonal forms of U may be obtained by permuting the relative positions of the eigenvalues along the diagonal. We choose R in such a way that all the eigenvalues 1 are in one submatrix, and -1 in the other, so that the matrix $\widetilde{U}$ can be represented in the form

$$\widetilde{U} = \begin{pmatrix} I & 0 \\ 0 & -I \end{pmatrix} = I \oplus (-I)$$

where I is the $2^{n-1} \times 2^{n-1}$ unit matrix and $\oplus$ is the direct sum. Since $\widetilde{V}^\pm$ commute with $\widetilde{U}$, they must have the forms

$$\widetilde{V}^\pm = \begin{pmatrix} \mathcal{A}^\pm & 0 \\ 0 & \mathcal{B}^\pm \end{pmatrix} = \mathcal{A}^\pm \oplus \mathcal{B}^\pm$$

where $\mathcal{A}^\pm$ and $\mathcal{B}^\pm$ are $2^{n-1} \times 2^{n-1}$ matrices. They are not necessarily diagonal. The $2^n \times 2^n$ matrix

$$\frac{1}{2}(I+\widetilde{U})$$

annihilates the lower submatrix and

$$\frac{1}{2}(I-\widetilde{U})$$

annihilates the upper submatrix. This means

$$\frac{1}{2}(I+\widetilde{U})\widetilde{V}^{+}=\begin{pmatrix} \mathcal{A}^{+} & 0 \\ 0 & 0 \end{pmatrix}, \qquad \frac{1}{2}(I-\widetilde{U})\widetilde{V}^{-}=\begin{pmatrix} 0 & 0 \\ 0 & \mathcal{B}^{-} \end{pmatrix}.$$

Therefore

$$\widetilde{V}=\begin{pmatrix} \mathcal{A}^{+} & 0 \\ 0 & \mathcal{B}^{-} \end{pmatrix}=\mathcal{A}^{+}\oplus\mathcal{B}^{-}.$$

To diagonalize V, it is sufficient to diagonalize $\widetilde{V}$, which has the same eigenvalues as V. Moreover, to diagonalize V it is sufficient to diagonalize $\mathcal{A}^{+}$ and $\mathcal{B}^{-}$ separately. The combined set of their eigenvalues constitutes the set of eigenvalues of V. We first diagonalize $\widetilde{V}^{+}$ and $\widetilde{V}^{-}$ separately and independently, thereby obtaining twice too many eigenvalues for each. To find the eigenvalues of $\frac{1}{2}(I+\widetilde{U})\widetilde{V}^{+}$ and $\frac{1}{2}(I-\widetilde{U})\widetilde{V}^{-}$, we might then decide which eigenvalues so obtained are to be discarded. This last step will not be necessary, however, for we show that as $n\to\infty$ a knowledge of the eigenvalues of $\widetilde{V}^{+}$ and $\widetilde{V}^{-}$ suffices to determine the largest eigenvalue of V. The set of eigenvalues of $\widetilde{V}^{\pm}$, however, is respectively equal to the set of eigenvalues of $V^{\pm}$. Therefore we diagonalize V^{+} and V^{-} separately and independently.

To find the eigenvalues of the matrices V^{+} and V^{-} we first find the eigenvalues of the rotations, of which V^{+} and V^{-} are spin representatives. These rotations are denoted by Ω^{+} and Ω^{-}, respectively which are both $2n\times 2n$ matrices. Thus we have the one-to-one mapping

$$V^{\pm}\leftrightarrow\Omega^{\pm}.$$

We find that

$$\Omega^{\pm}=\omega(1,2n|\mp 2i\phi)\left(\prod_{\alpha=1}^{n-1}\omega(2\alpha+1,2\alpha|-2i\phi)\right)\left(\prod_{\lambda=1}^{n}\omega(2\lambda,2\lambda-1|-2i\theta)\right)$$

where

$$\omega(\mu\nu|\alpha)=\omega(\nu\mu|-\alpha)$$

is the plane rotation in the plane $\mu\nu$ through the angle α. The eigenvalues of $\Omega^{\pm}$ are the same as that of

$$\omega^{\pm}=\Delta\Omega^{\pm}\Delta^{-1}$$

where Δ is the square root

$$\Delta := \sqrt{\prod_{\lambda=1}^{n} \omega(2\lambda, 2\lambda-1| - 2i\theta)} = \prod_{\lambda=1}^{n} \omega(2\lambda, 2\lambda-1| - i\theta).$$

Thus

$$\begin{aligned} \omega^{\pm} &= \Delta\chi^{\pm}\Delta \\ \Delta &= \omega(12|i\theta)\omega(34|i\theta)\cdots\omega(2n-1, 2n|i\theta) \\ \chi^{\pm} &= \omega(1, 2n| \pm 2i\phi)\left(\omega(23|2i\phi)\omega(45|2i\phi)\cdots\omega(2n-2, 2n-1|2i\phi)\right) \end{aligned}$$

We have

$$\Delta = J \oplus J \oplus \ldots \oplus J$$

where $\oplus$ is the direct sum and J is the 2×2 matrix

$$J = \begin{pmatrix} \cosh\theta & i\sinh\theta \\ -i\sinh\theta & \cosh\theta \end{pmatrix}$$

and

$$\chi^{\pm} = \begin{pmatrix} a & 0 & 0 & \cdots & & & \pm b \\ 0 & K & & & & & \\ 0 & & K & & & & \\ \vdots & & & & \ddots & & \vdots \\ & & & & & & 0 \\ & & & & & K & 0 \\ \mp b & & & & \cdots 0 & 0 & a \end{pmatrix}$$

where

$$K := \begin{pmatrix} \cosh 2\phi & i\sinh 2\phi \\ -i\sinh 2\phi & \cosh 2\phi \end{pmatrix}$$

and

$$a := \cosh(2\phi), \qquad b = i\sinh(2\phi).$$

Consequently, we obtain

$$\omega^{\pm} = \Delta\chi^{\pm}\Delta = \begin{pmatrix} A & B & 0 & 0 & \cdots & 0 & \mp B^* \\ B^* & A & B & 0 & & 0 & 0 \\ 0 & B^* & A & B & & & \vdots \\ \vdots & & & & & & \\ 0 & 0 & & & & & \\ \mp B & 0 & & & \cdots & B^* & A \end{pmatrix}$$

where A and B are 2×2 matrices given by

$$A := \begin{pmatrix} \cosh 2\phi \cosh 2\theta & -i \cosh 2\phi \sinh 2\theta \\ i \cosh 2\phi \sinh 2\theta & \cosh 2\phi \cosh 2\theta \end{pmatrix}$$

$$B := \begin{pmatrix} -\frac{1}{2} \sinh 2\phi \sinh 2\theta & i \sinh 2\phi \sinh^2 \theta \\ -i \sinh 2\phi \cosh^2 \theta & -\frac{1}{2} \sinh 2\phi \sinh 2\theta \end{pmatrix}.$$

To find the eigenvalues of $\omega^{\pm}$, we make the following ansatz for an eigenvector of $\omega^{\pm}$

$$\boldsymbol{\psi} = \begin{pmatrix} z\mathbf{u} \\ z^2\mathbf{u} \\ \vdots \\ z^n\mathbf{u} \end{pmatrix}$$

where z is a nonzero complex number and $\mathbf{u}$ is a two-component vector

$$\mathbf{u} = \begin{pmatrix} u_1 \\ u_2 \end{pmatrix}.$$

From the eigenvalue equation

$$\omega^{\pm}\boldsymbol{\psi} = \lambda\boldsymbol{\psi}$$

we obtain the following eigenvalue equations

$$\begin{aligned}
(zA + z^2B \mp z^nB^*)\mathbf{u} &= z\lambda\mathbf{u} \\
(z^2A + z^3B + zB^*)\mathbf{u} &= z^2\lambda\mathbf{u} \\
(z^3A + z^4B + z^2B^*)\mathbf{u} &= z^4\lambda\mathbf{u} \\
&\vdots \\
(z^{n-1}A + z^nB + z^{n-2}B^*)\mathbf{u} &= z^{n-1}\lambda\mathbf{u} \\
(z^nA \mp zB + z^{n-1}B^*)\mathbf{u} &= z^n\lambda\mathbf{u}.
\end{aligned}$$

There are only three independent eigenvalue equations

$$\begin{aligned}
(A + zB \mp z^{n-1}B^*)\mathbf{u} &= \lambda\mathbf{u} \\
(A + zB + z^{-1}B^*)\mathbf{u} &= \lambda\mathbf{u} \\
(A \mp z^{1-n}B + z^{-1}B^*)\mathbf{u} &= \lambda\mathbf{u}.
\end{aligned}$$

These eigenvalue equations are solved by setting

$$z^n = \mp 1.$$

These three equations then become the same one, namely

$$(A + zB + z^{-1}B^*)\mathbf{u} = \lambda\mathbf{u}$$

where the sign $\mp$ is associated with $\omega^\pm$. Thus, for ω^+ and for ω^-, there are n values of z, namely

$$z_k = \exp(2i\pi k/n), \qquad k = 0, 1, \dots, 2n-1$$

where

$$k = 1, 3, 5, \dots, 2n-1, \qquad \text{for } \omega^+$$
$$k = 0, 2, 4, \dots, 2n-2, \qquad \text{for } \omega^-.$$

For each k, two eigenvalues λ_k are determined by the equation

$$(A + z_k B + z_k^{-1} B^*)\mathbf{u} = \lambda_k \mathbf{u}$$

and λ_k is to be associated with $\omega^\pm$. This determines $2n$ eigenvalues each for $\omega^\pm$. To find λ_k, we note that

$$\det(A + z_k B + z_k^{-1} B^*) = 1.$$

Since the determinant of a 2×2 matrix is the product of the two eigenvalues we find that the two values of λ_k are given by

$$\lambda_k = \exp(\pm\gamma_k), \qquad k = 0, 1, \dots, 2n-1.$$

The two values of γ_k may be found from the equation

$$\frac{1}{2}\mathrm{tr}(A + z_k B + z_k^{-1} B^*) = \frac{1}{2}(\exp(\gamma_k) + \exp(-\gamma_k)) \equiv \cosh\gamma_k$$

where we have used the fact that the trace of a 2×2 matrix is the sum of the eigenvalues. Evaluating the trace we obtain

$$\cosh\gamma_k = \cosh 2\phi \cosh 2\theta - \cos\left(\frac{\pi k}{n}\right) \sinh 2\phi \sinh 2\theta$$

where $k = 0, 1, \dots, 2n-1$. If γ_k is a solution, then $-\gamma_k$ is also a solution. This possibility has already been taken into account. Therefore we define γ_k to be the positive solution. It can be shown that

$$\gamma_k = \gamma_{2n-k}, \qquad 0 < \gamma_0 < \gamma_1 < \cdots < \gamma_n.$$

The inequality $0 < \gamma_0 < \gamma_1 < \ldots < \gamma_n$ can be seen by noting that

$$\frac{\partial\gamma_k}{\partial k} = \frac{\pi}{n}\frac{\sin(\pi k/n)}{\sin\gamma_k}$$

which is positive for $k \le n$.

The eigenvalues of $\Omega^\pm$ are the same as those of $\omega^\pm$, respectively. Therefore $\Omega^\pm$ are products of commuting plane rotations. The 2^n eigenvalues of $V^\pm$ can now be written down with using lemma 3

$$\text{eigenvalues of } V^+ \text{ are } \exp\left(\frac{1}{2}(\pm\gamma_0 \pm \gamma_2 \pm \gamma_4 \pm \cdots \pm \gamma_{2n-2})\right)$$

$$\text{eigenvalues of } V^- \text{ are } \exp\left(\frac{1}{2}(\pm\gamma_1 \pm \gamma_3 \pm \gamma_5 \pm \cdots \pm \gamma_{2n-1})\right)$$

where all possible choices of the signs $\pm$ are to be made independently. Next we derive the eigenvalues of V. The set of eigenvalues of V consists of one-half the set of eigenvalues of V^+ and one-half that of V^-. The eigenvalues of $V^\pm$ are all positive and of order $\exp(n)$. Therefore all eigenvalues of V are positive and of order $\exp(n)$. We are only interested in the largest eigenvalue of V.

Let F and G be invertible $2^n \times 2^n$ matrices such that

$$F\left(\frac{1}{2}(I+\widetilde{U})\widetilde{V}^+\right)F^{-1} = V_D^+, \qquad G\left(\frac{1}{2}(I-\widetilde{U})\widetilde{V}^-\right)G^{-1} = V_D^-$$

where $V_D^\pm$ are diagonal matrices with half the eigenvalues of

$$\exp((\pm\gamma_0 \pm \gamma_2 \pm \ldots \pm \gamma_{2n-2})/2)$$

and

$$\exp((\pm\gamma_1 \pm \gamma_3 \pm \ldots \pm \gamma_{2n-1})/2)$$

respectively, appearing along the diagonal. It is possible to choose F and G in such a way that $F\widetilde{U}F^{-1}$ and $G\widetilde{U}\,G^{-1}$ remain diagonal matrices. Then F and G merely permute the eigenvalues of $\widetilde{U}$ along the diagonal. The convention has been adopted that $\widetilde{U}$ is given by $I \oplus (-I)$. Hence F and G either leave $\widetilde{U}$ unchanged or simply interchange the two sub-matrices I and $-I$. That is, F and G either commute or anticommute with $\widetilde{U}$. It follows that

$$V_D^+ = \frac{1}{2}(I \pm \widetilde{U})F\widetilde{V}^+F^{-1}, \qquad V_D^- = \frac{1}{2}(I \pm \widetilde{U})G\widetilde{V}^-G^{-1}$$

where the signs $\pm$ can be definitely determined by an explicit calculation. Now, we may write

$$\frac{1}{2}(I \pm \widetilde{U}) = \frac{1}{2}(I \pm \sigma_{z,1}\sigma_{z,2}\cdots\sigma_{z,n})$$

$$F\widetilde{V}^+F^{-1} = \prod_{k=1}^{n}\exp\left(\frac{1}{2}\gamma_{2k-1}\sigma_{z,P_k}\right)$$

$$G\widetilde{V}^-G^{-1} = \prod_{k=1}^{n}\exp\left(\frac{1}{2}\gamma_{2k-2}\sigma_{z,Q_k}\right)$$

where P and Q are two definite permutations of the integers $1, 2, \ldots, n$. The permutation P maps k into P_k and Q maps k into Q_k. The matrices $F\widetilde{V}^+F^{-1}$ and $G\widetilde{V}^-G^{-1}$ are arrived at by noting that they must respectively have the same eigenvalues as V^+ and V^-, except for possible different orderings of the eigenvalues. Since the eigenvalues of $\sigma_{z,k}$ are ± 1, we have

$$\frac{1}{2}(I \pm \widetilde{U}) = \begin{cases} 1 & \text{if an even number of } \sigma_{z,k} \text{ are } \pm 1 \\ 0 & \text{if an odd number of } \sigma_{z,k} \text{ are } \pm 1 \end{cases}$$

This condition is invariant under any permutation that maps $\sigma_{z,k}$ into σ_{z,P_k}. Therefore the eigenvalues of V_D^+ consists of those eigenvalues $\exp((\pm\gamma_0 \pm \gamma_2 \pm \ldots \pm \gamma_{2n-2})/2)$ for which an even (odd) number of $-$ signs appears in the exponents, if the $+(-)$ sign is chosen in $\frac{1}{2}(I \pm \widetilde{V})F\widetilde{V}^+F^{-1}$. We conclude that the

$$\text{largest eigenvalue of } V_D^+ \text{ is } \exp\left(\frac{1}{2}(\pm\gamma_0 + \gamma_2 + \gamma_4 + \cdots + \gamma_{2n-2})\right)$$

where the $\pm$ sign corresponds to the $\pm$ sign in

$$\frac{1}{2}(I \pm \widetilde{V})F\widetilde{V}^+F^{-1}.$$

As $n \to \infty$, these two possibilities give the same result, since γ_0 is negligible compared to the entire exponent. A similar conclusion can be found for V_D^-. Therefore we conclude that as $n \to \infty$ the

$$\text{largest eigenvalue of } V_D^+ = \exp\left(\frac{1}{2}(\gamma_0 + \gamma_2 + \gamma_4 + \cdots + \gamma_{2n-2})\right)$$

$$\text{largest eigenvalue of } V_D^- = \exp\left(\frac{1}{2}(\gamma_1 + \gamma_3 + \gamma_5 + \cdots + \gamma_{2n-1})\right).$$

Then the largest eigenvalue of V is given by

$$\Lambda = \exp\left(\frac{1}{2}(\gamma_1 + \gamma_3 + \gamma_5 + \cdots + \gamma_{2n-1})\right)$$

since $\gamma_k = \gamma_{2n-k}$ and $0 < \gamma_0 < \gamma_1 < \ldots < \gamma_n$. Now we evaluate explicitly the largest eigenvalue of V. We define

$$\mathcal{L} := \lim_{n\to\infty} \frac{1}{n} \ln \Lambda = \lim_{n\to\infty} \frac{1}{2n}(\gamma_1 + \gamma_2 + \gamma_5 + \cdots + \gamma_{2n-1}).$$

Let

$$\gamma(\nu) \equiv \gamma_{2k-1}, \qquad \nu := \frac{\pi}{n}(2k-1).$$

As $n \to \infty$, ν becomes a continuous variable, and we have

$$\sum_{k=1}^{n} \gamma_{2k-1} \to \frac{n}{2\pi} \int_0^{2\pi} d\nu \gamma(\nu).$$

Therefore

$$\mathcal{L} = \frac{1}{4\pi} \int_0^{2\pi} d\nu \gamma(\nu) = \frac{1}{2\pi} \int_0^{\pi} d\nu \gamma(\nu)$$

where we have used that $\gamma_k = \gamma_{n-k}$. Thus

$$\gamma(\nu) = \gamma(2\pi - \nu).$$

We note that $\gamma(\nu)$ is the positive solution of the equation

$$\cosh \gamma(\nu) = \cosh 2\phi \cosh 2\theta - \cos \nu \sinh 2\phi \sinh 2\theta$$

with

$$\phi = \beta J, \quad J > 0, \qquad \theta = \tanh^{-1}(\exp(-2\phi).$$

Since

$$\sinh(2\theta) = \frac{1}{\sinh(2\phi)}, \qquad \cosh(2\theta) = \coth(2\phi)$$

we find that

$$\cosh \gamma(\nu) = \cosh 2\phi \coth 2\phi - \cos \nu.$$

In the following we use the identity

$$|z| = \frac{1}{\pi} \int_0^\pi dt \ln(2 \cosh z - 2 \cos t).$$

Then we find that $\gamma(\nu)$ has the integral representation

$$\gamma(\nu) = \frac{1}{\pi} \int_0^\pi d\nu' \ln(2 \cosh 2\phi \coth 2\phi - 2 \cos \nu - 2 \cos \nu').$$

Therefore

$$\mathcal{L} = \frac{1}{2\pi^2} \int_0^\pi d\nu \int_0^\pi d\nu' \ln[2 \cosh 2\phi \coth 2\phi - 2(\cos \nu + \cos \nu')].$$

The integral remains unchanged if we let the region of integration be the rectangle

$$0 \leq \frac{\nu + \nu'}{2} \leq \pi, \qquad 0 \leq \nu - \nu' \leq \pi.$$

Let

$$\delta_1 \equiv \frac{\nu + \nu'}{2}, \qquad \delta_2 \equiv \nu - \nu'.$$

Then

$$\begin{aligned} \mathcal{L} &= \frac{1}{2\pi^2} \int_0^\pi d\delta_1 \int_0^\pi d\delta_2 \ln(2 \cosh 2\phi \coth 2\phi - 4 \cos \delta_1 \cos 2\delta_2) \\ &= \frac{1}{\pi} \int_0^{\pi/2} d\delta_2 \ln(2 \cos \delta_2) + \frac{1}{\pi} \int_0^{\pi/2} d\delta_2 \cosh^{-1} \left(\frac{D}{2 \cos \delta_2} \right) \end{aligned}$$

where

$$D := \cosh(2\phi)\coth(2\phi)\,.$$

Since $\cosh^{-1} x \equiv \ln[x + \sqrt{x^2 - 1}]$ we can write

$$\mathcal{L} = \frac{1}{2\pi}\int_0^\pi d\delta \ln[D(1 + \sqrt{1 - \kappa^2 \cos^2 \delta})]$$

where

$$\kappa = \frac{2}{D} = \frac{\exp(2\beta J) - \exp(-2\beta J)}{(\exp(2\beta J) + \exp(-2\beta J))^2}.$$

In the last integral $\cos^2 \delta$ may be replaced by $\sin^2 \delta$ without altering the value of the integral. Therefore

$$\mathcal{L} = \frac{1}{2}\ln\left(\frac{2\cosh^2 2\beta J}{\sinh 2\beta J}\right) + \frac{1}{2\pi}\int_0^\pi d\phi \ln\left(\frac{1}{2}(1 + \sqrt{1 - \kappa^2 \sin^2 \phi})\right).$$

Since

$$\mathcal{L} := \lim_{n\to\infty} \frac{1}{n}\ln \Lambda$$

and

$$\lim_{N\to\infty} \frac{1}{N}\ln Z(\beta) = \frac{1}{2}\ln(2\sinh(2\beta J)) + \lim_{n\to\infty}\frac{1}{n}\ln\Lambda$$

we find the partition function per lattice site. The Helmholtz free energy is given by

$$F(\beta) = -\frac{1}{\beta}\ln Z(\beta).$$

Exercises. (1) Show that the matrix P is symmetric.

(2) Show that for $\alpha \neq \beta$

$$[\sigma_{x,\alpha}, \sigma_{x,\beta}] = [\sigma_{y,\alpha}, \sigma_{y,\beta}] = [\sigma_{z,\alpha}, \sigma_{z,\beta}] = 0$$
$$[\sigma_{x,\alpha}, \sigma_{y,\beta}] = [\sigma_{x,\alpha}, \sigma_{z,\beta}] = [\sigma_{y,\alpha}, \sigma_{z,\beta}] = 0.$$

(3) Let X be an $n \times n$ matrix with $X^2 = I$ and $\theta \in \mathbf{R}$. Show that

$$\exp(\theta X) \equiv I \cosh\theta + X \sinh\theta.$$

(4) Show that

$$\exp(\theta\sigma_x) \otimes \exp(\theta\sigma_x \otimes \cdots \otimes \exp(\theta\sigma_x) \equiv \exp(\theta\sigma_{x,1})\exp(\theta\sigma_{x,2})\cdots\exp(\theta\sigma_{x,n})$$
$$\equiv \exp(\theta(\sigma_{x,1} + \sigma_{x,2} + \cdots + \sigma_{x,n})).$$

(5) Show that the matrices $\{\Gamma_\mu\}$ have the following properties:
(a) The dimensionality of Γ_μ cannot be smaller than $2^n \times 2^n$.
(b) If $\{\Gamma_\mu\}$ and $\{\Gamma'_\mu\}$ are two sets of matrices satisfying $\Gamma_\mu\Gamma_\nu + \Gamma_\nu\Gamma_\mu = 2\delta_{\nu\mu}$, there exists a nonsingular matrix S such that

$$\Gamma_\mu = S\Gamma_{\mu'}S^{-1}.$$

Show that the converse is also true.
(c) Any $2^n \times 2^n$ matrix is a linear combination of the unit matrix, the matrices Γ_μ (chosen to be $2^n \times 2^n$), and all the independent products $\Gamma_\mu\Gamma_\nu, \Gamma_\mu\Gamma_\nu\Gamma_l, \ldots$.

(6) Show that the set $\{\Gamma'_\mu\}$ satisfies

$$\Gamma'_\mu\Gamma'_\nu + \Gamma'_\nu\Gamma'_\mu = 2\delta_{\mu,\nu}.$$

(7) Show that

$$\omega(\mu\nu|\theta) = \omega(\nu\mu| - \theta), \qquad \omega^T(\mu\nu|\theta)\omega(\mu\nu|\theta) = I.$$

(8) Show that

$$U^2 = I, \qquad U(I+U) = I+U, \qquad U(I-U) = -(I-U)$$

$$U = i^n\Gamma_1\Gamma_2\cdots\Gamma_{2n}.$$

(9) Show that U commutes with a product of an even number of Γ_μ and anticommutes with a product of an odd number of Γ_μ.

(10) Show that the three matrices U, V^+ and V^- commute with one another.

(11) Discuss γ_k as a function of ϕ for $n \to \infty$.

3.6 One-Dimensional Isotropic Heisenberg Model

In this section we study the one-dimensional Heisenberg model with cyclic boundary conditions and the Yang Baxter relation. The Yang Baxter relation plays a central role in the investigation of exactly solvable models in statistical physics. The Yang Baxter relation includes the Kronecker product of matrix valued square matrices. We follow in our representation Takhtadzhyan and Faddeev [41], Kulish and Sklyanin [20], Baxter [7], Sogo and Wadati [31], Barouch [5].

The one-dimensional isotropic Heisenberg model describes a system of N interacting particles with spin 1/2 on a one-dimensional lattice. Let σ_j $j = 1, 2, 3$ be the Pauli spin matrices

$$\sigma_1 := \begin{pmatrix} 0 & 1 \\ 1 & 0 \end{pmatrix}, \qquad \sigma_2 := \begin{pmatrix} 0 & -i \\ i & 0 \end{pmatrix}, \qquad \sigma_3 := \begin{pmatrix} 1 & 0 \\ 0 & -1 \end{pmatrix}.$$

Let I_2 be the 2×2 unit matrix. We set

$$\sigma_{j,n} := I_2 \otimes \ldots \otimes I_2 \otimes \sigma_j \otimes I_2 \ldots \otimes I_2, \qquad j = 1, 2, 3$$

where σ_j stands at the n-th place and $n = 1, 2, \ldots, N$. Then the Hamilton operator (*isotropic Heisenberg model*) is defined as

$$\hat{H}_N := \frac{J}{4} \sum_{n=1}^{N} (\sigma_{1,n}\sigma_{1,n+1} + \sigma_{2,n}\sigma_{2,n+1} + \sigma_{3,n}\sigma_{3,n+1} - I)$$

where I is the $2^N \times 2^N$ unit matrix. In the following we omit the index N. Consequently, the operator $\hat{H}$ is a $2^N \times 2^N$ Hermitian matrix. Thus the underlying vector space is

$$\mathcal{H} = \prod_{n=1}^{N} \otimes \eta_n, \qquad \eta_n \equiv \mathbf{C}^2$$

with $\dim \mathcal{H} = 2^N$. We assume periodic boundary conditions

$$\sigma_{j,N+1} = \sigma_{j,1}$$

where $j = 1, 2, 3$.

Depending on the sign of the exchange constant J we distinguish the ferromagnetic case $J < 0$ and the antiferromagnetic case $J > 0$. The problem of most interest is to find the eigenvectors and eigenvalues of the Hamilton operator $\hat{H}$ and to investigate their asymptotic behaviour as $N \to \infty$.

The equations of motion are given by the *Heisenberg equation of motion.* Let $\hat{A}$ be an operator ($2^N \times 2^N$ matrix). The equations of motion are given by the Heisenberg equation of motion, i. e.

$$i\hbar \frac{d\hat{A}(t)}{dt} = [\hat{A}, \hat{H}](t)$$

where [,] denotes the commutator.

For the one-dimensional isotropic Heisenberg model there exists a so-called *local transition matrix*

$$L_n(\lambda) := \begin{pmatrix} \lambda I + \frac{i}{2}\sigma_{3,n} & \frac{i}{2}\sigma_{-,n} \\ \frac{i}{2}\sigma_{+,n} & \lambda I - \frac{i}{2}\sigma_{3,n} \end{pmatrix}$$

where

$$\sigma_{+,n} := \sigma_{1,n} + i\sigma_{2,n}$$

$$\sigma_{-,n} := \sigma_{1,n} - i\sigma_{2,n}$$

and $n = 1, 2, \ldots, N$. The entries of the 2×2 matrix L_n are $2^N \times 2^N$ matrices, i.e I is the $2^N \times 2^N$ unit matrix. Here λ is a so-called *spectral parameter.* Thus L_n is a matrix valued 2×2 matrix.

The importance of the local transition matrix lies in the fact that the constants of motion can be found with the help of the L_n. We describe this fact later.

Definition. A $2^N \times 2^N$ matrix $\hat{A}$ is called a *constant of motion* (for the isotropic Heisenberg model $\hat{H}$) if

$$[\hat{A}, \hat{H}] = 0\,.$$

Let

$$S_{1,N} := \frac{1}{2}\sum_{j=1}^{N}\sigma_{1,j}, \qquad S_{2,N} := \frac{1}{2}\sum_{j=1}^{N}\sigma_{2,j}, \qquad S_{3,N} := \frac{1}{2}\sum_{j=1}^{N}\sigma_{3,j}.$$

Furthermore the Heisenberg equation of motion for $S_{1,N}$, $S_{2,N}$, $S_{3,N}$ can be expressed with the help of L_n as follows. Consider the linear problem

$$L_n\psi_n = \psi_{n+1}, \qquad M_n\psi_n = \frac{d}{dt}\psi_n$$

where $n = 1, 2, \ldots, N$. Then the *compatability condition* (Steeb [38]) leads to the equation of motion

$$\left(\frac{d}{dt}L_n\right)\psi_n = \left(M_{n+1}L_n - L_nM_n\right)\psi_n$$

or as an operator equation

$$\frac{d}{dt}L_n = M_{n+1}L_n - L_nM_n.$$

The construction of the matrix-valued 2×2 matrix M_n is rather cumbersome. Here we refer to Barouch [5].

In order to introduce the Yang-Baxter relation we need the following two definitions. First we define the multiplication of an $n \times n$ matrix over $\mathbf{C}$ with a matrix valued $n \times n$ matrix.

Definition. Let A be an $n \times n$ matrix, where $a_{jk} \in \mathbf{C}$. Let B be an $n \times n$ matrix, where the entries B_{jk} are $m \times m$ matrices. Then we define

$$AB = C$$

where

$$C_{jk} = \sum_{l=1}^{n} a_{jl}B_{lk}.$$

Thus C is a matrix valued $n \times n$ matrix.

Furthermore we have to define the Kronecker product of two matrix valued 2×2 matrices.

Definition. Let A and B be 2×2 matrices where the entries A_{jk} and B_{jk}, respectively are $m \times m$ matrices. We define

$$A \otimes B := \begin{pmatrix} A_{11}B & A_{12}B \\ A_{21}B & A_{22}B \end{pmatrix}$$

where

$$A_{jk}B = \begin{pmatrix} A_{jk}B_{11} & A_{jk}B_{12} \\ A_{jk}B_{21} & A_{jk}B_{22} \end{pmatrix}.$$

Thus $A \otimes B$ is a matrix valued 4×4 matrix. We say that the Kronecker product is taken in the *auxiliary space*.

The matrix $L_n(\lambda)$ can also be written in the form (exercise (2))

$$L_n(\lambda) = \lambda I_2 \otimes I + \frac{i}{2}\sum_{j=1}^{3} \sigma_j \otimes \sigma_{j,n}$$

where I is the $2^N \times 2^N$ unit matrix.

Theorem. Let

$$R(\lambda)=\frac{1}{\lambda+i}\left(\left(\frac{\lambda}{2}+i\right)I_2\otimes I_2+\frac{\lambda}{2}\sum_{j=1}^{3}\sigma_j\otimes\sigma_j\right).$$

Then

$$R(\lambda-\mu)\,(L_n(\lambda)\otimes L_n(\mu))=(L_n(\mu)\otimes L_n(\lambda))\,R(\lambda-\mu)$$

holds, where the Kronecker product is taken in the auxiliary space. This is the famous *Yang-Baxter relation* (also known as the factorization equation or star-triangle equation). Thus $R(\lambda)$ is a c-number matrix, i. e. a 4×4 matrix over $\mathbf{C}$. The proof of this theorem is by straightforward calculation.

The matrix R can be written in the form

$$R(\lambda)=\begin{pmatrix}1&0&0&0\\0&b(\lambda)&c(\lambda)&0\\0&c(\lambda)&b(\lambda)&0\\0&0&0&1\end{pmatrix}$$

where

$$b(\lambda):=\frac{i}{\lambda+i},\qquad c(\lambda):=\frac{\lambda}{\lambda+i}.$$

The Yang Baxter relation is a sufficient condition for a system to be integrable (Baxter [7]).

Definition. The *monodromy matrix* $\mathcal{T}_N(\lambda)$ is defined by

$$\mathcal{T}_N(\lambda):=L_N(\lambda)\dots L_1(\lambda)\equiv\prod_{n=1}^{\widehat{N}}L_n(\lambda).$$

Similar to the local transition matrix $L_n(\lambda)$ it satisfies the relation

$$R(\lambda-\mu)(\mathcal{T}_N(\lambda)\otimes\mathcal{T}_N(\mu))=(\mathcal{T}_N(\mu)\otimes\mathcal{T}_N(\lambda))R(\lambda-\mu)$$

since $L_n(\lambda)$'s with different n commute.

The matrix valued 2×2 matrix $\mathcal{T}(\lambda)$ can be written as

$$\mathcal{T}_N(\lambda)=\begin{pmatrix}A_N(\lambda)&B_N(\lambda)\\C_N(\lambda)&D_N(\lambda)\end{pmatrix}.$$

The $2^N\times 2^N$ matrices $A_N(\lambda),B_N(\lambda),C_N(\lambda)$, and $D_N(\lambda)$ act in the space $\mathcal{H}$. We set

$$T_N(\lambda):=A_N(\lambda)+D_N(\lambda)=\mathrm{tr}\mathcal{T}_N(\lambda).$$

The following commutation relations hold (exercise (4))

$$[T_N(\lambda), T_N(\mu)] = 0, \qquad [B_N(\lambda), B_N(\mu)] = 0$$

and

$$A_N(\lambda)B_N(\mu) = \frac{1}{c(\mu-\lambda)}B_N(\mu)A_N(\lambda) - \frac{b(\mu-\lambda)}{c(\lambda-\mu)}B_N(\lambda)A_N(\mu)$$

$$D_N(\lambda)B_N(\mu) = \frac{1}{c(\lambda-\mu)}B_N(\mu)D_N(\lambda) - \frac{b(\lambda-\mu)}{c(\lambda-\mu)}B_N(\lambda)D_N(\mu).$$

The family of commuting matrices $T_N(\lambda)$ contains the momentum operator $\hat{P}_N$ and the Hamilton operator $\hat{H}$. Putting $\lambda = i/2$ in $L_n(\lambda)$ and using the definition of the monodromy matrix we find that the matrix $i^{-N}T_N(i/2)$ is unitary and coincides with the *cyclic shift matrix*, i. e.

$$\exp(-i\hat{P}_N)\sigma_{j,n}\exp(i\hat{P}_N) = \sigma_{j,n+1}$$

where $j = 1, 2, 3$ and $n = 1, 2, \ldots, N$. Its eigenvalues are given by

$$\exp(i\pi j), \qquad 0 \le P_j < 2\pi, \qquad j = 1, \ldots, 2^N.$$

Therefore

$$\hat{P}_N = \frac{1}{i}\ln i^{-N}T_N\left(\frac{i}{2}\right).$$

The Hamilton operator $\hat{H}$ is obtained by expanding the $\mathcal{T}_N(\lambda)$ in the neighbourhood of the point $\lambda = i/2$ and taking into account that $\exp(-i\hat{P}_N)\sigma_{j,n}\exp(i\hat{P}_N) = \sigma_{j,n+1}$. We find

$$\hat{H} = \frac{iJ}{2}\frac{d}{d\lambda}\ln T_N(\lambda)|_{\lambda=\frac{i}{2}} - \frac{NJ}{2}I.$$

To find higher order constants of motion we have to expand $T_N(\lambda)$ with respect to λ.

Next we study states for the one-dimensional isotropic Heisenberg model $\hat{H}$. Consider the vectors

$$\boldsymbol{\omega}_n := \begin{pmatrix} 1 \\ 0 \end{pmatrix}, \quad n = 1, \ldots, N, \qquad \boldsymbol{\Omega}_N = \prod_{n=1}^{N} \otimes \boldsymbol{\omega}_n, \quad \boldsymbol{\Omega}_N \in \mathcal{H}.$$

In the following we omit the index N.

We find the following relations hold

$$A_N(\lambda)\boldsymbol{\Omega} = \left(\lambda + \frac{i}{2}\right)^N \boldsymbol{\Omega}, \qquad D_N(\lambda)\boldsymbol{\Omega} = \left(\lambda - \frac{i}{2}\right)^N \boldsymbol{\Omega}, \qquad C_N(\lambda)\boldsymbol{\Omega} = \mathbf{0}.$$

It follows that the vector

$$\mathbf{\Psi}_N(\lambda_1, \lambda_2, \ldots, \lambda_\ell) = B_N(\lambda_1)B_N(\lambda_2)\ldots B_N(\lambda_\ell)\mathbf{\Omega}$$

is an eigenvector of the family of operators $T_N(\lambda)$ if the complex numbers $\lambda_1, \ldots, \lambda_\ell$ satisfy the system of equations

$$\left(\frac{\lambda_j - \frac{i}{2}}{\lambda_j + \frac{i}{2}}\right) = \prod_{\substack{k=1\\k\neq j}}^{\ell} \frac{\lambda_j - \lambda_k - i}{\lambda_j - \lambda_k + i} \qquad j = 1, \ldots, \ell.$$

We call these eigenvectors *Bethe's vectors.* The corresponding eigenvalue $\Lambda(\lambda; \lambda_1, \ldots, \lambda_\ell)$ has the form

$$\Lambda(\lambda; \lambda_1, \ldots, \lambda_\ell) = \left(\lambda + \frac{i}{2}\right)^N \prod_{j=1}^{\ell} \frac{\lambda - \lambda_j - i}{\lambda - \lambda_j} + \left(\lambda - \frac{i}{2}\right)^N \prod_{j=1}^{\ell} \frac{\lambda - \lambda_j + i}{\lambda - \lambda_j}.$$

Since the matrices $B_N(\lambda)$ commute, i. e. $[B(\lambda), B(\mu)] = 0$ both the vector $\mathbf{\Psi}_N(\lambda_1, \ldots, \lambda_\ell)$ and the eigenvalue $\Lambda(\lambda;\ \lambda_1, \ldots, \lambda_\ell)$ are symmetric functions of $\lambda_1, \ldots, \lambda_\ell$.

The eigenvalues of the operators $\hat{P}_N$ and $\hat{H}_N$ have the form

$$p(\lambda_1, \ldots, \lambda_\ell) = \frac{1}{i} \sum_{j=1}^{\ell} \ln \frac{\lambda_j + \frac{i}{2}}{\lambda_j - \frac{i}{2}} \quad \text{modulo}\, 2\pi, \qquad h(\lambda_1, \ldots, \lambda_\ell) = -\frac{J}{2} \sum_{j=1}^{\ell} \frac{1}{\lambda_j^2 + \frac{1}{4}}.$$

In addition to the spectral parameter λ, it is convenient to introduce the variable $p(\lambda)$

$$\exp(ip(\lambda)) = \frac{\lambda + \frac{i}{2}}{\lambda - \frac{i}{2}}.$$

Thus

$$p(\lambda) = -2\mathrm{arctg}(2\lambda) + \pi \quad \text{modulo } 2\pi\,.$$

In the variables P_j the momentum p and the energy h have the form

$$\begin{aligned} p(\lambda_1, \ldots, \lambda_\ell) &= \sum_{j=1}^{\ell} P_j \quad \text{modulo } 2\pi \\ h(\lambda_1, \ldots, \lambda_\ell) &= -J \sum_{j=1}^{\ell} (1 - \cos P_j). \end{aligned}$$

The vector $\mathbf{\Omega}$ plays the role of a vacuum, and the matrix $B_N(\lambda)$ has the meaning of the creation operator with the momentum $p(\lambda)$ and the energy

$$h(\lambda) = \frac{J}{2} \frac{dp(\lambda)}{d\lambda} = -J(1 - \cos p(\lambda)).$$

A necessary condition for Bethe's vector not to vanish is

$$\ell \leq N/2.$$

We can calculate the nomalization of Bethe's vector using the commutation relation

$$[C_N(\lambda), B_N(\mu)] = \frac{b(\lambda-\mu)}{c(\lambda-\mu)}(A_N(\mu)D_N(\lambda) - A_N(\lambda)D_N(\mu)).$$

We can also prove the simpler assertion that Bethe's vectors with different collections $(\lambda_1, \ldots, \lambda_\ell)$, $(\lambda_1', \ldots, \lambda_{\ell'}')$ are orthogonal.

In addition to $\hat{P}$ and $\hat{H}$, among the observable values of our system, we have the spin operator

$$\hat{S}_j := \frac{1}{2}\sum_{n=1}^{N} \sigma_{j,n}, \qquad j = 1, 2, 3.$$

We omit the index N in the cases when this cannot cause misunderstanding.

The matrices $\hat{P}$ and $\hat{H}$ commute with $\hat{S}_j$. By straightforward calculation we find that the following relation holds

$$\hat{S}_+\mathbf{\Psi} = \mathbf{0}$$

where

$$\hat{S}_+ := \frac{1}{2}\sum_{n=1}^{N} \sigma_{+,n}, \qquad \hat{S}_- := \frac{1}{2}\sum_{n=1}^{N} \sigma_{-,n}\,.$$

For the isotropic Heisenberg model the Bethe's ansatz does not determine all eigenvectors of the Hamilton operator. Together with Bethe's vectors, vectors of the form

$$\hat{S}_-^m \mathbf{\Psi}$$

are also eigenvectors, where $1 \leq m \leq 2L$ and L is the spin of the representation to which $\mathbf{\Psi}$ belongs. Moreover, the eigenvalue equation

$$\hat{S}_3\mathbf{\Psi} = \left(\frac{N}{2} - \ell\right)\mathbf{\Psi}$$

holds. From the formula for the square of the spin S^2 with eigenvalues $L(L+1)$, $L \geq 0$

$$\hat{S}^2 = \sum_{j=1}^{3} \hat{S}_j^2 = \hat{S}_-\hat{S}_+ + \hat{S}_3(\hat{S}_3 + I)$$

it follows that for Bethe's vectors

$$L = \frac{N}{2} - \ell.$$

Therefore the important inequality follows

$$\ell \leq \frac{N}{2}.$$

We find that

$$[\hat{S}_j, T(\lambda)] = 0, \quad j = 1, 2, 3, \qquad [\hat{S}_3, B(\lambda)] = -B(\lambda), \qquad [\hat{S}_+, B(\lambda)] = A(\lambda) - D(\lambda).$$

By construction the vector $\mathbf{\Omega}$ is also an eigenvector for the operators $\hat{S}_+$, $\hat{S}_3$. We find

$$\hat{S}_+\mathbf{\Omega} = \mathbf{0}, \qquad \hat{S}_3\mathbf{\Omega} = \frac{N}{2}\mathbf{\Omega}.$$

We have, after carrying $\hat{S}_+$ through all operators $B(\lambda_j)$ to the vector $\mathbf{\Omega}$,

$$\hat{S}_+\mathbf{\Psi} = \sum_{j=1}^{\ell} B(\lambda_1)\dots B(\lambda_{j-1})(A(\lambda_j) - D(\lambda_j))B(\lambda_{j+1})\dots B(\lambda_\ell)\mathbf{\Omega}.$$

Using the permutation relations given above we carry the matrices $A(\lambda_j)$ and $D(\lambda_j)$ through the $B(\lambda_k)$ to the vector $\mathbf{\Omega}$. We arrive at

$$\hat{S}_+\mathbf{\Psi} = \sum_{j=1}^{\ell} M_j(\lambda_1, \dots, \lambda_\ell)B(\lambda_1)\dots B(\lambda_{j-1})B(\lambda_{j+1})\dots B(\lambda_\ell)\mathbf{\Omega}.$$

To obtain $M_1(\lambda_1, \dots, \lambda_\ell)$ we have to carry $A(\lambda_1) - D(\lambda_1)$ through the chain $B(\lambda_2)\dots B(\lambda_\ell)$ to the vector $\mathbf{\Omega}$. Therefore,

$$M_1(\lambda_1, \dots, \lambda_\ell) = \left(\lambda_1 + \frac{i}{2}\right)^N \prod_{j=2}^{\ell} \frac{\lambda_1 - \lambda_j - i}{\lambda_1 - \lambda_j} - \left(\lambda_1 - \frac{i}{2}\right)^N \prod_{j=2}^{\ell} \frac{\lambda_1 - \lambda_j + i}{\lambda_1 - \lambda_j}.$$

The remaining coefficients $M_j(\lambda_1 \dots, \lambda_\ell)$ are obtained from $M_1(\lambda_1, \dots, \lambda_\ell)$ by the corresponding permutation of the numbers $\lambda_1, \dots, \lambda_\ell$. They have the following form

$$M_j(\lambda_1, \dots, \lambda_\ell) = \left(\lambda_j + \frac{i}{2}\right)^N \prod_{\substack{k=1 \\ k\neq j}}^{\ell} \frac{\lambda_j - \lambda_k - i}{\lambda_j - \lambda_k} - \left(\lambda_j - \frac{i}{2}\right)^N \prod_{\substack{k=1 \\ k\neq j}}^{\ell} \frac{\lambda_j - \lambda_k + i}{\lambda_j - \lambda_k}$$

where $j = 1, \dots, \ell$. The system of equations

$$\left(\frac{\lambda_j - \frac{i}{2}}{\lambda_j + \frac{i}{2}}\right) = \prod_{\substack{k=1 \\ k\neq j}}^{\ell} \frac{\lambda_j - \lambda_k - i}{\lambda_j - \lambda_k + i}, \qquad j = 1, \dots, \ell$$

means exactly that

$$M_j(\lambda_1, \dots, \lambda_\ell) = 0, \qquad j = 1, \dots, \ell.$$

For further reading we refer to Takhtadzhyan and Faddeev [41], Kulish and Sklyanin [20], Baxter [7], Sogo and Wadati [31], Barouch [5].

Exercises. (1) Show that

$$\left(I \otimes I - \sum_{j=1}^{3} \sigma_j \otimes \sigma_j\right)^2 = 4\left(I \otimes I - \sum_{j=1}^{3} \sigma_j \otimes \sigma_j\right).$$

(2) Let $J < 0$. Show that $\hat{H}$ can be represented in the form

$$\hat{H} = -J \sum_{n=1}^{N} (\sigma_{1,n} \otimes \sigma_{1,n+1} + \sigma_{2,n} \otimes \sigma_{2,n+1} + \sigma_{3,n} \otimes \sigma_{3,n+1} - I)^2.$$

(3) Let

$$S_{N,1} := \frac{1}{2} \sum_{j=1}^{N} \sigma_{1,j}, \qquad S_{N,2} := \frac{1}{2} \sum_{j=1}^{N} \sigma_{2,j}, \qquad S_{N,3} := \frac{1}{2} \sum_{j=1}^{N} \sigma_{3,j}.$$

Find the equations of motion for $S_{N,1}$, $S_{N,2}$ and $S_{N,3}$.

(4) Show that the matrix $L_n(\lambda)$ can also be written in the form

$$L_n(\lambda) = \lambda I_2 \otimes I + \frac{i}{2} \sum_{j=1}^{3} \sigma_j \otimes \sigma_{j,n}$$

where I is the $2^N \times 2^N$ unit matrix.

(5) Show that R can be written in the form

$$R(\lambda) = \begin{pmatrix} 1 & 0 & 0 & 0 \\ 0 & b(\lambda) & c(\lambda) & 0 \\ 0 & c(\lambda) & b(\lambda) & 0 \\ 0 & 0 & 0 & 1 \end{pmatrix}$$

where

$$b(\lambda) := \frac{i}{\lambda + i}, \qquad c(\lambda) := \frac{\lambda}{\lambda + i}.$$

(6) Show that the following commutation relations hold

$$[T_N(\lambda), T_N(\mu)] = 0, \qquad [B_N(\lambda), B_N(\mu)] = 0$$

and

$$A_N(\lambda)B_N(\mu) = \frac{1}{c(\mu - \lambda)} B_N(\mu)A_N(\lambda) - \frac{b(\mu - \lambda)}{c(\lambda - \mu)} B_N(\lambda)A_N(\mu)$$

$$D_N(\lambda)B_N(\mu) = \frac{1}{c(\lambda - \mu)} B_N(\mu)D_N(\lambda) - \frac{b(\lambda - \mu)}{c(\lambda - \mu)} B_N(\lambda)D_N(\mu).$$

(7) Show that

$$\sigma_{j,n}\sigma_{k,n} = \delta_{jk} I + i \sum_{l=1}^{3} \epsilon_{jkl} \sigma_{l,n}$$

and

$$[\sigma_{j,m}, \sigma_{k,n}] = 2i \sum_{l=1}^{3} \epsilon_{jkl} \sigma_{l,m} \delta_{mn}$$

where

$$\epsilon_{jkl} := \begin{cases} 0 & \text{if two indices } jkl \text{ are the same} \\ 1 & \text{if } jkl \text{ is an even permutation of 1, 2, 3} \\ -1 & \text{if } jkl \text{ is an odd permutation of 1, 2, 3.} \end{cases}$$

(8) Show that the following relations hold

$$A_N(\lambda)\mathbf{\Omega} = \left(\lambda + \frac{i}{2}\right)^N \mathbf{\Omega}, \qquad D_N(\lambda)\mathbf{\Omega} = \left(\lambda - \frac{i}{2}\right)^N \mathbf{\Omega}, \qquad C_N(\lambda)\mathbf{\Omega} = \mathbf{0}.$$

(9) Show that the eigenvalues of the operators $\hat{P}_N$ and $\hat{H}_N$ have the form

$$p(\lambda_1, \ldots, \lambda_\ell) = \frac{1}{i} \sum_{j=1}^{\ell} \ln \frac{\lambda_j + \frac{i}{2}}{\lambda_j - \frac{i}{2}} \quad \text{modulo } 2\pi, \qquad h(\lambda_1, \ldots, \lambda_\ell) = -\frac{J}{2} \sum_{j=1}^{\ell} \frac{1}{\lambda_j^2 + \frac{1}{4}}.$$

(10) Show that the matrices $\hat{P}$ and $\hat{H}$ commute with $\hat{S}_j$.

(11) Show that

$$[\hat{S}_j, T(\lambda)] = 0, \quad j = 1, 2, 3, \qquad [\hat{S}_3, B(\lambda)] = -B(\lambda), \qquad [\hat{S}_+, B(\lambda)] = A(\lambda) - D(\lambda).$$

(12) Let $J < 0$. Show that the Hamilton operator $\hat{H}$ annihilates the vector $\mathbf{\Omega}$, i.e.

$$\hat{H}\mathbf{\Omega} = \mathbf{0}.$$

The vector $\mathbf{\Omega}$ is the ground state - the *ferromagnetic vacuum*.

3.7 Quantum Groups

First we introduce the concept of a *Hopf algebra* (Abe [1]). Let G be a finite group and K a field. The set $A = \text{Map}(G, K)$ of all functions defined on G with values in K becomes a K-algebra when we define the scalar product and the sum and product of functions by

$$(\alpha f)(x) = \alpha f(x), \qquad (f+g)(x) = f(x) + g(x)$$

$$(fg)(x) = f(x)g(x), \quad f, g \in A, \quad \alpha \in K, \; x \in G.$$

In general, a K-algebra A can be characterized as a K-linear space A together with two K-linear maps

$$\mu : A \otimes_K A \to A, \qquad \eta : K \to A$$

which satisfy axioms corresponding to the associative law and the unitary property respectively. If we identify $A \otimes_K A$ with $\text{Map}(G \times G, K)$ where $A = \text{Map}(G, K)$, and if the operations of G are employed in defining the K-linear maps

$$\Delta : A \to A \otimes_K A, \qquad \epsilon : A \to K$$

respectively by $\Delta f(x \otimes y) := f(xy)$, $\epsilon f = f(e)$ for $x, y \in G$ and where e is the identity element of the group G, then Δ and ϵ become homomorphisms of K-algebras having properties which are dual to μ and η, respectively. In general, a K-linear space A with K-linear maps $\mu, \eta, \Delta, \epsilon$ defined as above is called a K-bialgebra. Furthermore, we can define a k-linear endomorphism of $A = \text{Map}(G, K)$

$$S : A \to A, \qquad (Sf)(x) = f(x^{-1}), \quad f \in A, \quad x \in G$$

such that the equalities

$$\mu(1 \otimes S)\Delta = \mu(S \otimes 1)\Delta = \eta \circ \epsilon$$

hold. A K-bialgebra on which we can define a K-linear map S as above is called a K-Hopf algebra. Thus, a K-Hopf algebra is an algebraic system which simultaneously admits structures of a K-algebra as well as its dual, where these two structures are related by a certain specific law. For a finite group G, its group ring KG over a field K is the dual space of the k-linear space $A = \text{Map}(G, K)$ where its K-algebra structure is given by the dual K-linear maps of Δ and ϵ. Moreover, kG admits a K-Hopf algebra structure when we take the dual K-linear maps of μ, η, and S. In other words, KG is the dual K-Hopf algebra of $\text{Map}(G, K)$. If we replace the finite group G in the above argument by a topological group and K by the field of real numbers or the field of complex numbers, or if we take G to be an algebraic group over an algebraically closed field K and A is replaced by the K-algebra of all continuous representative functions or of all regular functions over G, then A turns out to be a K-Hopf algebra in exactly the same manner. These algebraic systems play an important role when studying the structure of G. Similarly, a K-Hopf algebra structure can be defined naturally on the universal enveloping algebra of a K-Lie algebra. The universal enveloping algebra of the Lie algebra of a semi-simple algebraic group turns out to be (in a sense) the dual of the Hopf algebra defined above.

Quantum groups and quantum algebras are remarkable mathematical structures which have found unexpected applications in theoretical physics, in particular statistical physics, in the last few years. For example, knot theory and quantum algebras are closely related. A quantum group is a non-commutative and non-cocommutative Hopf algebra. Algebraically, these structures are readily described as continuous deformations of the familiar Lie (super) algebras widely used in physics, a description which uses many ideas from classical q analysis. Two-dimensional statistical models, field theory and S-matrix theory can be described by the quantum Yang-Baxter equation, which applies quantum algebras. Thus the quantum Yang-Baxter equation plays an important role. With solutions of quantum Yang-Baxter equation one can construct exactly solvable models and find their eigenvalues and eigenstates. On the other hand, any solution of the quantum Yang-Baxter equation can be generally used to find the new quasi-triangular Hopf algebra. Many multiparameter solutions (4×4 matrices) of the quantum Yang-Baxter equation have been obtained. Corresponding to the case of the standard one-parameter R matrix, the algebras related to the standard two-parameter R matrix have also been discussed. We show how computer algebra can be used to investigate quantum algebras. This means we show with an example how the Kronecker product and the Yang-Baxter relation can be implemented using computer algebra. Furthermore we also show how the quantum algebra obtained from the Yang-Baxter relation can be implemented.

The starting point in the construction of the Yang-Baxter equation is the 2×2 matrix

$$T := \begin{pmatrix} a & b \\ c & d \end{pmatrix}$$

where a, b, c and d are noncommutative linear operators. We may consider a, b, c and d as $n \times n$ matrices over the complex or real numbers. In other words, T is a matrix-valued matrix. Let I be the 2×2 identity matrix

$$I := \begin{pmatrix} 1 & 0 \\ 0 & 1 \end{pmatrix}$$

where 1 is the unit operator (identity matrix). Now we define the 4×4 matrices

$$T_1 := T \otimes I, \qquad T_2 := I \otimes T$$

where $\otimes$ denotes the Kronecker product. Thus T_1 and T_2 are matrix (operator) valued 4×4 matrices. Applying the rules for the Kronecker product, we find

$$T_1 = T \otimes I = \begin{pmatrix} a & 0 & b & 0 \\ 0 & a & 0 & b \\ c & 0 & d & 0 \\ 0 & c & 0 & d \end{pmatrix}, \qquad T_2 = I \otimes T = \begin{pmatrix} a & b & 0 & 0 \\ c & d & 0 & 0 \\ 0 & 0 & a & b \\ 0 & 0 & c & d \end{pmatrix}.$$

The algebra related to this quantum matrix is governed by the *Yang-Baxter equation*

$$R_q T_1 T_2 = T_2 T_1 R_q$$

where R_q is an R-matrix and q is a nonzero complex number.

Let us now consider examples.

As an example consider now the 4×4 matrix

$$R_q := \begin{pmatrix} 1 & 0 & 0 & 0 \\ 0 & -1 & 0 & 0 \\ 0 & 1+q & q & 0 \\ 0 & 0 & 0 & 1 \end{pmatrix}$$

where q is a nonzero complex number. The Yang-Baxter equation gives rise to the relations of the algebra elements a, b, c and d

$$ab = q^{-1}ba, \quad dc = qcd, \quad bc = -qcb$$

$$bd = -db, \quad ac = -ca, \quad [a,d] = (1+q^{-1})bc$$

where all the commutative relations have been omitted. Here $[\,,\,]$ denotes the commutator. This quantum matrix T can be considered as a linear transformation of plane $\mathcal{A}_q(2)$ with coordinates (x, ξ) satisfying

$$x\xi = -\xi x.$$

It is straightforward to prove that the coordinate transformations deduced by T

$$\begin{pmatrix} x' \\ \xi' \end{pmatrix} = \begin{pmatrix} a & b \\ c & d \end{pmatrix} \begin{pmatrix} x \\ \xi \end{pmatrix}$$

keep the relation

$$x'\xi' = -\xi' x'.$$

As there is no nilpotent element in the quantum matrix T, there exists no constraint on the coordinates x and ξ. This quantum plane is the same as the one related to $GL_q(2)$. From the algebraic relations for a, b, c, and d we can define an element of the algebra,

$$\delta(T) \equiv \delta := ad - q^{-1}bc.$$

δ satisfies the following relations

$$[a, \delta] = 0, \qquad [d, \delta] = 0$$

$$\{c, \delta\}_q \equiv qc\delta + \delta c = 0, \qquad \{\delta, b\}_q \equiv q\delta b + b\delta = 0.$$

The element δ commutes only with a and b and hence is not the centre of the algebra.

If we consider the R-matrix

$$R_q := \begin{pmatrix} q & 0 & 0 & 0 \\ 0 & 1 & 0 & 0 \\ 0 & q-q^{-1} & 1 & 0 \\ 0 & 0 & 0 & q \end{pmatrix}$$

then we obtain from the Yang-Baxter equation the relations

$$ab = qba, \quad ac = qca, \quad bc = cb, \quad bd = qdb, \quad cd = qdc, \quad ad - da = (q - q^{-1})bc.$$

Next we give an implementation with the software package REDUCE. From a computational point of view two tasks have to be performed: The first task would be the implementation of the relations for a, b, c and d. Then these commutation relations should be used to evaluate a new commutation relation. An example is the definition of the operator δ and evaluation of the commutation relations for δ.

In our implementation for the evaluation of the relations of a, b, c and d we set: $a = a(1)$, $b = a(2)$, $c = a(3)$, $d = a(4)$ and $\delta = del$. The nonzero complex number q is denoted by q. The implementation of this task in REDUCE is as follows:

```
%gr2.red;

%a, del are noncommutative operators;
operator a, del, r1, r2, r3, r4;
noncom a, del, r1, r2, r3, r4;

% implementation of rules (7);
let a(1)*a(2) = a(2)*a(1)/q;
let a(1)*a(3) = -a(3)*a(1);
let a(1)*a(4) = a(4)*a(1) + (1+1/q)*a(2)*a(3);
let a(2)*a(3) = -q*a(3)*a(2);
let a(2)*a(4) = -a(4)*a(2);
let a(3)*a(4) = (1/q)*a(4)*a(3);

% definition of delta (see (9));
del := a(1)*a(4) - (1/q)*a(2)*a(3);

% evaluation of (10);
r1 := a(1)*del - del*a(1);
r2 := a(4)*del - del*a(4);
r3 := q*a(3)*del + del*a(3);
r4 := q*del*a(2) + a(2)*del;
```

The output shows that the commutation relations for δ are satisfied.

In our second task we have to evaluate the Yang-Baxter relation. In the program we implement the Kronecker product of $T \otimes I$ and $I \otimes T$ and the Yang-Baxter relation. The Kronecker product is implemented as a procedure. We have to take into account that a, b, c, d are noncommutative operators (matrices). We give the implementation in REDUCE.

```
%qg.red;

procedure Kron(A,B);
begin
n := 2; m := n*n;
operator A$  matrix AA(n,n)$
for i:=1:n do
for j:=1:n do
AA(i,j):=A(i,j)$
operator B$  matrix BB(n,n)$
for i:=1:n do
for j:=1:n do
BB(i,j):=B(i,j)$

operator C$  matrix CC(m,m);
c1 := 0; c2 := 0;
for r:=1:n do
for s:=1:n do
begin
for i:=1:n do
for j:=1:n do
begin
c1 := n*(r-1); c2 := n*(s-1);
CC(i+c1,j+c2) := AA(r,s)*BB(i,j);
end; end;
return CC;
end;       % end procedure kron;

operator a, b, c, d, id;  noncom a, b, c, d;

operator T;
matrix T1(4,4); matrix T2(4,4); matrix R(4,4);

T(1,1) := a(j); T(1,2) := b(j); T(2,1) := c(j); T(2,2) := d(j);
ID(1,1) := 1; ID(1,2) := 0; ID(2,1) := 0; ID(2,2) := 1;

T1 := Kron(T,ID);  T2 := Kron(ID,T);
```

```
R(1,1) := 1; R(1,2) := 0;   R(1,3) := 0; R(1,4) := 0;
R(2,1) := 0; R(2,2) := -1;  R(2,3) := 0; R(2,4) := 0;
R(3,1) := 0; R(3,2) := 1+q; R(3,3) := q; R(3,4) := 0;
R(4,1) := 0; R(4,2) := 0;   R(4,3) := 0; R(4,4) := 1;

matrix RES(4,4);

RES := R*T1*T2 - T2*T1*R;
```

The output is

```
t1 := mat((a(j),0,b(j),0),(0,a(j),0,b(j)),(c(j),0,d(j),0),
(0,c(j),0,d(j)))$
t2 := mat((a(j),b(j),0,0),(c(j),d(j),0,0),(0,0,a(j),b(j)),
(0,0,c(j),d(j)))$

res := mat((0, - a(j)*b(j)*q + b(j)*a(j),
- a(j)*b(j)*q + b(j)*a(j),0),( - (a(j)*c(j) + c(j)*a(j)),
- a(j)*d(j) - c(j)*b(j)*q - c(j)*b(j) + d(j)*a(j),
- (b(j)*c(j) + c(j)*b(j)*q), - (b(j)*d(j) + d(j)*b(j))),
(q*(a(j)*c(j) + c(j)*a(j)),b(j)*c(j) + c(j)*b(j)*q,
- a(j)*d(j)*q + b(j)*c(j)*q + b(j)*c(j) + d(j)*a(j)*q,
q*(b(j)*d(j) + d(j)*b(j))),(0, - c(j)*d(j)*q + d(j)*c(j),
- c(j)*d(j)*q + d(j)*c(j),0))$
```

Thus we find the relations given by the relations for a, b, c, and d since $res = 0$.

The program can now easily be extended to higher dimensions. In higher dimensions the evaluations become very lengthy. If T is a 3×3 matrix, then $I_3 \otimes T$ and $T \otimes I_3$ are 9×9 matrices. Then the Yang-Baxter relation yields 81 equations (of course some of them are identically zero). Here the application of computer algebra is very helpful.

A comprehensive survey on quantum algebras is given by Zachos [43].

3.8 Lax Representation

In this section we show how the Kronecker product can be used to find new Lax representations.

A number of interesting dynamical systems can be written in *Lax representation* (Steeb [36], Steeb [38], Steeb and Lai [39])

$$\frac{dL}{dt} = [A, L](t)$$

where A and L are given by $n \times n$ matrices. The time-dependent matrices A and L are called a *Lax pair*. An example is the Toda lattice. Given two Lax representations we show how the Kronecker product can be used to find a new Lax representation. We then give an application. Finally, we discuss some extensions.

Let X and R be $m \times m$ matrices and Y and P be $n \times n$ matrices. Then we have

$$(X \otimes Y)(R \otimes P) \equiv (XR) \otimes (YP).$$

This identity will be used in the following.

Theorem. Given two Lax representations

$$\frac{dL}{dt} = [A, L](t), \qquad \frac{dM}{dt} = [B, M](t)$$

where L and A are $m \times m$ matrices and M and B are $n \times n$ matrices. Let I_m be the $m \times m$ unit matrix and I_n be the $n \times n$ unit matrix. Then we find the Lax representation

$$\frac{d}{dt}(L \otimes M) = [A \otimes I_n + I_m \otimes B, L \otimes M](t).$$

We call this the Kronecker product Lax representation.

Proof. This can be seen as follows: The right hand side can be written as

$$[A \otimes I_n + I_m \otimes B, L \otimes M] = [A \otimes I_n, L \otimes M] + [I_m \otimes B, L \otimes M].$$

Thus we find that

$$[A \otimes I_n + I_n \otimes B, L \otimes M] = (AL) \otimes M - (LA) \otimes M + L \otimes (BM) - L \otimes (MB).$$

On the other hand we have

$$\frac{d}{dt}(L \otimes M) = \frac{dL}{dt} \otimes M + L \otimes \frac{dM}{dt}.$$

Inserting $dL/dt = [A, L](t)$ and $dM/dt = [B, M](t)$ into this equation completes the proof. ♠

First integrals of $dL/dt = [A, L](t)$ can be found from

$$F_k = \mathrm{tr}(L^k), \qquad k = 1, 2, \ldots \quad .$$

Since

$$\mathrm{tr}(X^k \otimes Y^j) = \mathrm{tr}(X^k)\mathrm{tr}(Y^j)$$

where X is an $m \times m$ matrix and Y is an $n \times n$ matrix, we find that first integrals of the Kronecker product Lax representation are given by

$$F_{kj} = \mathrm{tr}(L^k)\mathrm{tr}(M^j), \qquad k, j = 1, 2, \ldots \quad .$$

Obviously, we can extend this to more than two Lax pairs. For example, given three Lax representations

$$\frac{dL}{dt} = [A, L](t), \qquad \frac{dM}{dt} = [B, M](t), \qquad \frac{dN}{dt} = [C, N](t)$$

we find that

$$\frac{d}{dt}(L \otimes M \otimes N) = [A \otimes I_n \otimes I_p + I_m \otimes B \otimes I_p + I_m \otimes I_n \otimes C, L \otimes M \otimes N](t)$$

where C and N are $p \times p$ matrices.

Example. Consider the nonlinear system of ordinary differential equations

$$\frac{du_1}{dt} = (\lambda_3 - \lambda_2)u_2u_3, \qquad \frac{du_2}{dt} = (\lambda_1 - \lambda_3)u_3u_1, \qquad \frac{du_3}{dt} = (\lambda_2 - \lambda_1)u_1u_2$$

where $\lambda_j \in \mathbf{R}$. This system describes Euler's rigid body motion. The first integrals are given by

$$I_1(\mathbf{u}) = u_1^2 + u_2^2 + u_3^2, \qquad I_2(\mathbf{u}) = \lambda_1 u_1^2 + \lambda_2 u_2^2 + \lambda_3 u_3^2.$$

A Lax representation is given by

$$\frac{dL}{dt} = [L, \lambda L](t)$$

where

$$L := \begin{pmatrix} 0 & -u_3 & u_2 \\ u_3 & 0 & -u_1 \\ -u_2 & u_1 & 0 \end{pmatrix}, \qquad \lambda L := \begin{pmatrix} 0 & -\lambda_3 u_3 & \lambda_2 u_2 \\ \lambda_3 u_3 & 0 & -\lambda_1 u_1 \\ -\lambda_2 u_2 & \lambda_1 u_1 & 0 \end{pmatrix}.$$

Then $\mathrm{tr}(L)^k$ $(k = 1, 2, \ldots)$ provides only one first integral. We obtain

$$\mathrm{tr}L = 0, \qquad \mathrm{tr}L^2 = -2(u_1^2 + u_2^2 + u_3^2) = -2I_1.$$

Since L does not depend on λ we cannot find I_2. This Lax representation can be applied to the product Lax representation when we set $M = L$. ♣

To overcome the problem of finding only one first integral we consider now

$$\frac{d(L+Ay)}{dt} = [L+Ay, \lambda L + By](t),$$

where y is a dummy variable and A and B are time-independent diagonal matrices, i.e., $A = \text{diag}\,(A_1, A_2, A_3)$ and $B = \text{diag}\,(B_1, B_2, B_3)$ with $A_j, B_j \in \mathbf{R}$. The equation decomposes into various powers of y, namely

$$\begin{aligned} y^0 &: \frac{dL}{dt} = [L, \lambda L] \\ y^1 &: 0 = [L, B] + [A, \lambda L] \\ y^2 &: [A, B] = 0. \end{aligned}$$

The last equation is satisfied identically since A and B are diagonal matrices. The second equation leads to

$$\lambda_i = \frac{B_j - B_k}{A_j - A_k}$$

where (i, j, k) are permutations of (1, 2, 3). It can be satisfied by setting

$$B_j = A_j^2$$

and

$$\lambda_i = A_j + A_k.$$

Consequently the original Lax pair $L, \lambda L$ satisfies the extended Lax pair

$$L + Ay, \qquad \lambda L + By.$$

Now $\text{tr}[(L+Ay)^2]$ and $\text{tr}[(L+Ay)^3]$ provide both first integrals given above. For this extended Lax pair the concept of the Kronecker product described above can also be applied.

For some dynamical systems such as the energy level motion we find an extended Lax representation

$$\frac{dL}{dt} = [A, L](t), \qquad \frac{dK}{dt} = [A, K](t)$$

where L and K do not commute. For this extended system of Lax representations we can also apply the Kronecker product technique given above in order to find new Lax representations.

3.9 Signal Processing

The development of a Fast Fourier transform algorithm has made a significant impact on the field of digital signal processing. Motivated by this success, many researchers have since developed fast transform implementations corresponding to a wide variety of discrete unitary transforms, and at present fast algorithms are known for a variety of unitary transforms such as Hadamard, Haar, Slant, Discrete Cosine, and Hartley transforms, etc. (see Regalia and Mitra [29] and Elliott and Rao [13] and reference therein). Each unitary transform family has its own characteristic properties depending on its basis functions, which in some sense match a transform to an application. For example, the discrete Fourier transform is well suited to frequency domain analysis and filtering, the Discrete Cosine transform to data compression, the Slant transform to image coding, the Hadamard and Haar transforms to dyadic-invariant signal processing, and these and others to generalized spectral analysis. The suitablility of a unitary transform in a given application depends not only on its basis functions, but also on the existence of an efficient computational algorithm. Fast transform algorithms are typically developed by recognizing various patterns of the elements of a discrete unitary transform matrix. The existence of such patterns implies some redundancy among the matrix elements which can be exploited in developing sparse matrix factorizations. The product of sparse matrices can result in a much simplified computational algorithm compared to the direct implementation of a matrix equation.

The utility of Kronecker products as a compact notation for representing matrix patterns has been recognized by various researchers (see Regalia and Mitra [29] and Elliott and Rao [13] and references therein). The Kronecker product representations lead to efficient computer implementations for numerous discrete unitary transforms and Kronecker products can be defined in terms of matrix factorizations and play a central role in generalized spectral analysis. A wide class of discrete unitary transforms can be generated using recursion formulas of generalized Kronecker products with matrix permutations (i.e., element reordering). Many fast transform flowgraphs may be obtained as structural interpretations of their matrix notation. Thus the Kronecker product decomposition of various unitary transforms plays a central role in developing fast transform algorithms.

Regalia and Mitra [29] proposed a generalization of the Kronecker product and described its utility with some examples from the field of signal processing. A large class of discrete unitary transforms can be developed from a single recursion formula. Closed-form expressions are derived for sparse matrix factorizations in terms of this generalized matrix product. When applied to discrete unitary transform matrices, fast transform algorithms can be developed directly upon recognizing patterns in the matrices, without resorting to tedious matrix factorization steps. They also derived some apparently novel properties of Hadamard transformations and polyadic permutations in the context of Kronecker products. They also showed the invariance of Hadamard matrices under a bit-permuted ordering similarity transformation, and established a simple result of the Kronecker decomposability of any polyadic permutation matrix. Closed form expressions relating this matrix product to sparse matrix factorizations are obtained, which are particularly useful in factorizing "patterned" matrices. Many unitary matrices, for example, are patterned matrices, for which the sparse matrix factorizating equivalent of this matrix product directly yields fast transform algorithms, without resorting to the tedious matrix factorization steps. They also describe a fast transform algorithm to developing equivalent filter bank representations.

We start with the connection of the Kronecker product and the *bit-reversed order*. Let

$$a, b, c, d, e, f \in \{0, 1\}.$$

Consider the Kronecker product of the vectors

$$\begin{pmatrix} a \\ b \end{pmatrix} \otimes \begin{pmatrix} c \\ d \end{pmatrix} \otimes \begin{pmatrix} e \\ f \end{pmatrix} = \begin{pmatrix} ace \\ acf \\ ade \\ adf \\ bce \\ bcf \\ bde \\ bdf \end{pmatrix}.$$

On the other hand

$$\begin{pmatrix} b \\ a \end{pmatrix} \otimes \begin{pmatrix} d \\ c \end{pmatrix} \otimes \begin{pmatrix} f \\ e \end{pmatrix} = \begin{pmatrix} bdf \\ bde \\ bcf \\ bce \\ adf \\ ade \\ acf \\ ace \end{pmatrix}.$$

Thus we find the bit-reversed order.

Regalia and Mitra [29] introduced the Kronecker product of an $m \times n$ matrix A and a $k \times l$ matrix B as the $mk \times nl$ matrix

$$A \otimes B := \begin{pmatrix} Ab_{00} & Ab_{01} & \dots & Ab_{0,l-1} \\ Ab_{10} & Ab_{11} & \dots & Ab_{1,l-1} \\ \vdots & \vdots & & \vdots \\ Ab_{k-1,0} & Ab_{k-1,1} & \dots & Ab_{k-1,l-1} \end{pmatrix}.$$

We recall that this definition of $A \otimes B$ can be transformed into the definition given in chapter 2 using permutation matrices. Furthermore in the definition given above we count from zero, which is useful when we look at software implementation in C-C++.

The generalization of the definition of the Kronecker product is as follows:

Definition. Given a set of N $(m \times r)$ matrices A_i, $i = 0, 1, \cdots, N-1$, denoted by $\{A\}_N$, and a $(N \times l)$ matrix B, we define the $(mN \times rl)$ matrix $(\{A\}_N \otimes B)$ as

$$\{A\}_N \otimes B := \begin{pmatrix} A_0 \otimes \mathbf{b}_0 \\ A_1 \otimes \mathbf{b}_1 \\ \vdots \\ A_{N-1} \otimes \mathbf{b}_{N-1} \end{pmatrix}$$

where $\mathbf{b}_i$ denotes the ith row vector of the matrix B. If each matrix A_i is identical, then the definition reduces to the usual Kronecker product of matrices.

Example. Let

$$\{A\}_2 = \left\{ \begin{matrix} \begin{pmatrix} 1 & 1 \\ 1 & -1 \end{pmatrix} \\ \\ \begin{pmatrix} 1 & -i \\ 1 & i \end{pmatrix} \end{matrix} \right\}, \qquad B = \begin{pmatrix} 1 & 1 \\ 1 & -1 \end{pmatrix}.$$

From the definition we obtain

$$\{A\}_2 \otimes B = \begin{pmatrix} \begin{pmatrix} 1 & 1 \\ 1 & -1 \end{pmatrix} & \otimes & (1 \quad 1) \\ \\ \begin{pmatrix} 1 & -i \\ 1 & i \end{pmatrix} & \otimes & (1 \quad -1) \end{pmatrix} = \begin{pmatrix} 1 & 1 & 1 & 1 \\ 1 & -1 & 1 & -1 \\ 1 & -i & -1 & i \\ 1 & i & -1 & -i \end{pmatrix}$$

which is a (4×4) discrete Fourier transform matrix with the rows arranged in bit-reversed order (see section 2.9). ♣

In the definition and the example we have assumed that the number of matrices in the set $\{A\}_N$ matches the number of rows in the matrix B. B need not be a single matrix, but may also be a periodic sequence of $(k \times l)$ matrices (and hence a periodic sequence of row vectors).

The matrix product $\{A\} \otimes \{B\}$ is obtained as the Kronecker product of each matrix in $\{A\}$ with each row vector in $\{B\}$, and is easily verified to yield a periodic sequence of $(mk \times rl)$ matrices, which thus admits a finite representation in one period. To affix notation, a set of matrices will always be indicated using the bracket notation (e.g., $\{A\}$), whereas if the set consists of a single matrix, the brackets will be omitted.

Example. Let

$$\{A\} = \left\{ \begin{array}{c} \begin{pmatrix} 1 & 1 \\ 1 & -1 \end{pmatrix} \\ \\ \begin{pmatrix} 1 & -i \\ 1 & i \end{pmatrix} \\ \\ \begin{pmatrix} 1 & e^{-i\pi/4} \\ 1 & -e^{-i\pi/4} \end{pmatrix} \\ \\ \begin{pmatrix} 1 & e^{-i3\pi/4} \\ 1 & -e^{-i3\pi/4} \end{pmatrix} \end{array} \right\}, \qquad \{B\} = \left\{ \begin{array}{c} \begin{pmatrix} 1 & 1 \\ 1 & -1 \end{pmatrix} \\ \\ \begin{pmatrix} 1 & -i \\ 1 & i \end{pmatrix} \end{array} \right\}.$$

Then

$$\{A\} \otimes \{B\} = \left\{ \begin{array}{c} \begin{pmatrix} 1 & 1 & 1 & 1 \\ 1 & -1 & 1 & -1 \\ 1 & -i & -1 & i \\ 1 & i & -1 & -i \end{pmatrix} \\ \\ \begin{pmatrix} 1 & e^{-i\pi/4} & -i & e^{-i3\pi/4} \\ 1 & e^{-i\pi/4} & -i & -e^{-i3\pi/4} \\ 1 & e^{-i3\pi/4} & i & e^{-i\pi/4} \\ 1 & -e^{-i3\pi/4} & i & -e^{-i\pi/4} \end{pmatrix} \end{array} \right\}.$$

Each matrix in the resulting set has dimensions (4×4). ♣

By comparing the examples we observe that this matrix product can result in a single matrix only if the rightmost multiplicand is a single matrix. The following algebraic property of this generalized Kronecker product is straightforward to prove

$$(\{A\} \otimes \{B\}) \otimes \{C\} = \{A\} \otimes (\{B\} \otimes \{C\}).$$

Definition. Let $\{A\}_N$ be a sequence of $(m \times n)$ matrices and E be a single $(n \times r)$ matrix. Then

$$\{A\}E := \left\{ \begin{array}{c} A_0 E \\ A_1 E \\ \vdots \\ A_{N-1} E \end{array} \right\}$$

where each matrix in the sequence in $(m \times r)$.

This leads to the identity

$$(\{A\}E) \otimes (\{B\}F) \equiv (\{A\} \otimes \{B\})(E \otimes F).$$

The next two identities are useful in developing sparse matrix factorizations.

$$\{A\}_N \otimes I_N = \bigoplus_{i=0}^{N-1} A_i$$

where $\oplus$ denotes the direct product. The proof is easy to see by noting that both sides of the equation yield a block-diagonal matrix containing the matrices A_i.

Let p sets of matrices be denoted by $\{A^{(k)}\}_{N_k}$, $k = 0, 1, \cdots, p-1$, where each matrix is $(m \times n)$, and the kth set has $N_k = m^k$ matrices. Consider the matrix R defined by

$$R := \{A^{(p-1)}\}_{m^{p-1}} \otimes \{A^{(p-2)}\}_{m^{(p-2)}} \otimes \cdots \otimes \{A^{(1)}\}_m \otimes A^{(0)}.$$

Note that, by construction, the last matrix $A^{(0)}$ is a single matrix, as is R. The matrix R admits the sparse matrix factorization

$$R = \prod_{k=0}^{p-1} \left(\bigoplus_{i=0}^{m^{p-k-1}-1} \left(I_{n^k} \otimes A_i^{(p-k-1)} \right) \right).$$

If each matrix $A_i^{(k)}$ is (para-)unitary, then R is (para-) unitary. This result follows from the fact that the Kronecker product, direct sum, or matrix product of (para-)unitary matrices is a (para-) unitary matrix.

If $m = n$, such that each matrix $A_i^{(k)}$, $i = 0, \cdots, m^k - 1$, $k = 0, \ldots, p-1$, is square, then

$$\det R = \prod_{k=0}^{p-1} \prod_{i=0}^{m^k-1} (\det A_i^{(k)})^{m^k}.$$

This relation can be found by successive application of the algebra of determinants.

Example. We find a fast transform algorithm for an (8×8) complex *BIFORE matrix*. This matrix is given by

$$R_8 := \begin{pmatrix} 1 & 1 & 1 & 1 & 1 & 1 & 1 & 1 \\ 1 & -1 & 1 & -1 & 1 & -1 & 1 & -1 \\ 1 & -i & 1 & i & 1 & -i & -1 & i \\ 1 & i & -1 & -i & 1 & i & -1 & -i \\ 1 & 1 & -i & -i & -1 & -1 & i & i \\ 1 & -1 & -i & i & -1 & 1 & i & -i \\ 1 & 1 & i & i & -1 & -1 & -i & -i \\ 1 & -1 & i & -i & -1 & 1 & -i & i \end{pmatrix}.$$

The matrix can be partitioned as

$$R_8 = \begin{pmatrix} \begin{pmatrix} 1 & 1 \\ 1 & -1 \end{pmatrix} \otimes (1 \quad 1 \quad 1 \quad 1) \\ \\ \begin{pmatrix} 1 & -i \\ 1 & i \end{pmatrix} \otimes (1 \quad -1 \quad 1 \quad -1) \\ \\ \begin{pmatrix} 1 & 1 \\ 1 & -1 \end{pmatrix} \otimes (1 \quad -i \quad -1 \quad i) \\ \\ \begin{pmatrix} 1 & 1 \\ 1 & -1 \end{pmatrix} \otimes (1 \quad i \quad -1 \quad -i) \end{pmatrix}.$$

Thus we can write

$$R_8 = \left\{ \begin{matrix} \begin{pmatrix} 1 & 1 \\ 1 & -1 \end{pmatrix} \\ \\ \begin{pmatrix} 1 & -i \\ 1 & i \end{pmatrix} \\ \\ \begin{pmatrix} 1 & 1 \\ 1 & -1 \end{pmatrix} \\ \\ \begin{pmatrix} 1 & 1 \\ 1 & -1 \end{pmatrix} \end{matrix} \right\} \otimes \begin{pmatrix} 1 & 1 & 1 & 1 \\ 1 & -1 & 1 & -1 \\ 1 & -i & -1 & i \\ 1 & i & -1 & -i \end{pmatrix}.$$

Note that the (4×4) matrix on the right-hand side is a 4-point complex BIFORE transform. Denoting this matrix by R_4 and continuing the process, we obtain

$$R_4 = \begin{pmatrix} 1 & 1 & 1 & 1 \\ 1 & -1 & 1 & -1 \\ 1 & -i & -1 & i \\ 1 & i & -1 & -i \end{pmatrix} = \left\{ \begin{matrix} \begin{pmatrix} 1 & 1 \\ 1 & -1 \end{pmatrix} \\ \\ \begin{pmatrix} 1 & -i \\ 1 & i \end{pmatrix} \end{matrix} \right\} \otimes \begin{pmatrix} 1 & 1 \\ 1 & -1 \end{pmatrix}.$$

Thus we obtain the sparse matrix factorization

$$R_8 = \left(\begin{pmatrix} 1 & 1 \\ 1 & -1 \end{pmatrix} \oplus \begin{pmatrix} 1 & -i \\ 1 & i \end{pmatrix} \oplus \begin{pmatrix} 1 & 1 \\ 1 & -1 \end{pmatrix} \oplus \begin{pmatrix} 1 & 1 \\ 1 & -1 \end{pmatrix} \right)$$

$$\times \begin{pmatrix} I_2 \otimes \begin{pmatrix} 1 & 1 \\ 1 & -1 \end{pmatrix} & \\ & I_2 \otimes \begin{pmatrix} 1 & -i \\ 1 & i \end{pmatrix} \end{pmatrix} \left(I_4 \otimes \begin{pmatrix} 1 & 1 \\ 1 & -1 \end{pmatrix} \right). \quad \clubsuit$$

We find that a recursion formula for the family of complex BIRORE matrices is easily expressed in the present notation. In particular, we have

$$R_N = \{B\}_{N/2} \otimes R_{N/2}, \qquad R_1 = 1$$

where N is a power of two, and $\{B\}_{N/2}$ has $N/2$ matrices in the set, with the ith matrix (counting form $i = 0$ to $N/2 - 1$) given by

$$B_i = \begin{cases} \begin{pmatrix} 1 & -i \\ 1 & i \end{pmatrix}, & i = 1 \\ \\ \begin{pmatrix} 1 & 1 \\ 1 & -1 \end{pmatrix}, & \text{otherwise.} \end{cases}$$

Discrete Fourier transform matrices may also be expressed using the recursion relation given above provided the rows are permuted into bit-reversed order (which, in a practical implementation, requires the output samples to be sorted if they are desired in normal order). In this case the ith matrix of $\{B\}_{N/2}$ is given by

$$B_i = \begin{pmatrix} 1 & W_N^{\ll i \gg} \\ 1 & -W_N^{\ll i \gg} \end{pmatrix}, \qquad i = 0, 1, \cdots, N/2 - 1$$

where

$$W_N := \exp(-i2\pi/N)$$

and $\ll i \gg$ is the decimal number obtained from a bit reversal of a $\log_2(N/2)$-bit binary representation of i. The matrix sets in the example given above are recognized as $\{B\}_4$ and $\{B\}_2$, for example. Although the recursion relation corresponds to radix-2 Fast Fourier Transform algorithms, a similar recursion can be developed for any prime radix. For example, to obtain a radix 3 Fast Fourier Transform recursion $N/2$ can be replaced everywhere with $N/3$, and the i-th matrix in $\{B\}_{N/3}$ can be defined as

$$B_i = \begin{pmatrix} 1 & W_N^{\ll i \gg} & W_N^{2\ll i \gg} \\ 1 & \exp(i4\pi/3)W_N^{\ll i \gg} & \exp(i2\pi/3)W_N^{2\ll i \gg} \\ 1 & \exp(i2\pi/3)W_N^{\ll i \gg} & \exp(i4\pi/3)W_N^{2\ll i \gg} \end{pmatrix}, \qquad i = 0, 1, \cdots, N/3 - 1$$

where now $\ll i \gg$ is the decimal number obtained from a trit reversal of a $\log_3(n/3)$-trit ternary representation of i, and so on. Nonprime radices can be split into prime factors, each of which may be treated in a manner analogous to that above.

Modified Walsh-Hadamard Transform matrices may also be expressed with the recursion given above. In this case the ith matrix of $\{B\}_{N/2}$ is given by

$$B_i = \begin{cases} \begin{pmatrix} 1 & 1 \\ 1 & -1 \end{pmatrix}, & i = 0 \\ \\ \sqrt{2}I_2, & \text{otherwise.} \end{cases}$$

We point out that the Haar transform may be derived from the modified Walsh-Hadamard transform through zonal bit-reversed ordering permutations of the input and output data.

Hadamard matrices are known to satisfy the recursion

$$R_N = \begin{pmatrix} 1 & 1 \\ 1 & -1 \end{pmatrix} \otimes R_{N/2}, \qquad R_1 = 1$$

which may be understood as a special case of the recursion given above in which each B_i is identical.

With the recursion given above, the matrix B_N in all cases defined above satisfies

$$\widetilde{R}_N R_N = R_N \widetilde{R}_N = N I_N$$

so that $R_N/\sqrt{N}$ is unitary. Therefore fast transform algorithms in all cases are immediately available.

Regalia and Mitra [29] also describe filter bank applications using the definitions and recursions given above. Moreover, they describe matrix factorization of a polyadic permutation matrix with application to a programmable polyadic shift structure.

Chapter 4

Tensor Product

4.1 Hilbert Spaces

In this section we introduce the concept of a Hilbert space. Hilbert spaces play the central rôle in quantum mechanics. The proofs of the theorems given in this chapter can be found in Prugovečki [28]. We assume that the reader is familiar with the notation of a linear space. First we introduce the pre-Hilbert space.

Definition. A linear space L is called a *pre-Hilbert space* if there is defined a numerical function called the *scalar product* (or *inner product*) which assigns to every f, g of "vectors" of L $(f, g \in L)$ a complex number. The scalar product satisfies the conditions

$$(a) \qquad (f,f) \geq 0; \qquad (f,f) = 0 \quad \text{iff} \quad f = 0$$

$$(b) \qquad (f,g) = \overline{(g,f)}$$

$$(c) \qquad (cf,g) = c(f,g) \text{ where } c \text{ is an arbitrary complex number}$$

$$(d) \qquad (f_1 + f_2, g) = (f_1, g) + (f_2, g)$$

where $\overline{(g,f)}$ denotes the complex conjugate of (g,f).

It follows that

$$(f, g_1 + g_2) = (f, g_1) + (f, g_2)$$

and

$$(f, cg) = \bar{c}(f, g).$$

Definition. A linear space E is called *normed space*, if for every $f \in E$ there is associated a real number $||f||$, the norm of the vector f such that

$$(a) \qquad ||f|| \geq 0 \quad ; \quad ||f|| = 0 \quad \text{iff} \quad f = 0$$

$$(b) \qquad ||cf|| = |c|||f|| \text{ where } c \text{ is an arbitrary complex number}$$

$$(c) \qquad ||f + g|| \leq ||f|| + ||g||.$$

The topology of a normed linear space E is thus defined by the distance

$$d(f, g) := ||f - g||.$$

If a scalar product is given we can introduce a norm. The norm of f is defined by

$$||f|| := \sqrt{(f, f)}.$$

A vector $f \in L$ is called normalized if $||f|| = 1$.

Let $f, g \in L$. The following identity holds (*parallelogram identity*)

$$||f + g||^2 + ||f - g||^2 = 2(||f||^2 + ||g||^2).$$

Definition. Two functions $f \in L$ and $g \in L$ are called *orthogonal* if

$$(f, g) = 0.$$

Definition. A sequence $\{f_n\}$ ($n \in \mathbf{N}$) of elements in a normed space E is called a *Cauchy sequence* if, for every $\epsilon > 0$, there exists a number M_ϵ such that $||f_p - f_q|| < \epsilon$ for $p, q > M_\epsilon$.

Definition. A normed space E is said to be *complete* if every Cauchy sequence of elements in E converges to an element in E.

Example. Let $\mathbf{Q}$ be the rational numbers. Since the sum and product of two rational numbers are again rational numbers we obviously have a pre-Hilbert space with the scalar product $(q_1, q_2) := q_1 q_2$. However, the pre-Hilbert space is not complete. Consider the sequence

$$f_n = 1 + \frac{1}{1!} + \frac{1}{2!} + \ldots + \frac{1}{(n-1)!}$$

with $n = 1, 2, \ldots$. The sequence f_n is obviously a Cauchy sequence. However

$$\lim_{n \to \infty} f_n \to e$$

and $e \notin \mathbf{Q}$. ♣

Definition. A complete pre-Hilbert space is called a *Hilbert space.*

Definition. A complete normed space is called a *Banach space.*

A Hilbert space will be denoted by $\mathcal{H}$ in the following. A Banach space will be denoted by $\mathcal{B}$ in the following.

Theorem. Every pre-Hilbert space L admits a completion $\mathcal{H}$ which is a Hilbert space.

Example. Let $L = \mathbf{Q}$. Then $\mathcal{H} = \mathbf{R}$. ♣

Before we discuss some examples of Hilbert spaces we give the definitions of strong and weak convergence in Hilbert spaces.

Definition. A sequence $\{f_n\}$ of vectors in a Hilbert space $\mathcal{H}$ is said to *converge strongly* to f if

$$\|f_n - f\| \to 0$$

as $n \to \infty$. We write $s - \lim_{n \to \infty} f_n \to f$.

Definition. A sequence $\{f_n\}$ of vectors in a Hilbert space $\mathcal{H}$ is said to *converge weakly* to f if

$$(f_n, g) \to (f, g)$$

as $n \to \infty$, for any vector g in $\mathcal{H}$. We write $w - \lim_{n \to \infty} f_n \to f$.

It can be shown that strong convergence implies weak convergence. The converse is not generally true, however.

Let us now give several examples of Hilbert spaces which are important in quantum mechanics.

Example 1. Every finite dimensional vector space with an inner product is a Hilbert space. Let $\mathbf{C}^n$ be the linear space of n-tuples of complex numbers with the scalar product

$$(\mathbf{u}, \mathbf{v}) := \sum_{j=1}^{n} u_j \bar{v}_j .$$

Then $\mathbf{C}^n$ is a Hilbert space. Let $\mathbf{u} \in \mathbf{C}^n$. We write the vector $\mathbf{u}$ as a column vector

$$\mathbf{u} = \begin{pmatrix} u_1 \\ u_2 \\ \cdot \\ \cdot \\ u_n \end{pmatrix}.$$

Thus we can write the scalar product in matrix notation

$$(\mathbf{u}, \mathbf{v}) = \mathbf{u}^T \bar{\mathbf{v}}$$

where $\mathbf{u}^T$ is the transpose of $\mathbf{u}$. ♣

Example 2. By $l_2(\mathbf{N})$ we mean the set of all infinite dimensional vectors (sequences) $\mathbf{u} = (u_1, u_2, \ldots)^T$ of complex numbers u_j such that

$$\sum_{j=1}^{\infty} |u_j|^2 < \infty.$$

Here $l_2(\mathbf{N})$ is a linear space with operations ($a \in \mathbf{C}$)

$$\begin{aligned} a\mathbf{u} &= (au_1, au_2, \ldots)^T \\ \mathbf{u} + \mathbf{v} &= (u_1 + v_1, u_2 + v_2, \ldots)^T \end{aligned}$$

with $\mathbf{v} = (v_1, v_2, \ldots)^T$ and $\sum_{j=1}^{\infty} |v_j|^2 < \infty$. One has

$$\sum_{j=1}^{\infty} |u_j + v_j|^2 \leq \sum_{j=1}^{\infty} (|u_j|^2 + |v_j|^2 + 2|u_j v_j|) \leq 2 \sum_{j=1}^{\infty} (|u_j|^2 + |v_j|^2) < \infty .$$

The scalar product is defined as

$$(\mathbf{u}, \mathbf{v}) := \sum_{j=1}^{\infty} u_j \bar{v}_j = \mathbf{u}^T \bar{\mathbf{v}}.$$

It can also be proved that this pre-Hilbert space is complete. Therefore $l_2(\mathbf{N})$ is a Hilbert space. As an example, let us consider

$$\mathbf{u} = (1, \frac{1}{2}, \frac{1}{3}, \ldots, \frac{1}{n}, \ldots)^T.$$

Since

$$\sum_{n=1}^{\infty} \frac{1}{n^2} < \infty$$

we find that $\mathbf{u} \in l_2(\mathbf{N})$. Let

$$\mathbf{u} = (1, \frac{1}{\sqrt{2}}, \frac{1}{\sqrt{3}}, \ldots, \frac{1}{\sqrt{n}}, \ldots)^T.$$

Then $\mathbf{u} \notin l_2(\mathbf{N})$. ♣

Example 3. $L_2(M)$ is the space of Lebesgue square-integrable functions on M, where M is a Lebesgue measurable subset of $\mathbf{R}^n$, where $n \in \mathbf{N}$. If $f \in L_2(M)$, then

$$\int_M |f|^2 \, dm < \infty.$$

The integration is performed in the Lebesgue sense. The scalar product in $L_2(M)$ is defined as

$$(f, g) := \int_M f(x)\bar{g}(x) \, dm$$

where $\bar{g}$ denotes the complex conjugate of g. It can be shown that this pre-Hilbert space is complete. Therefore $L_2(M)$ is a Hilbert space. Instead of dm we also write dx in the following. If the Riemann integral exists then it is equal to the Lebesgue integral. However, the Lebesgue integral exists also in cases in which the Riemann integral does not exist. ♣

Example 4. Consider the linear space M^n of all $n \times n$ matrices over $\mathbf{C}$. The *trace* of an $n \times n$ matrix $A = (a_{jk})$ is given by

$$\mathrm{tr}A := \sum_{j=1}^{n} a_{jj}.$$

We define a scalar product by

$$(A, B) := \mathrm{tr}(AB^*)$$

where tr denotes the trace and B^* denotes the conjugate transpose matrix of B. We recall that $\mathrm{tr}(C + D) = \mathrm{tr}C + \mathrm{tr}D$ where C and D are $n \times n$ matrices. ♣

Example 5. Consider the linear space of all infinite dimensional matrices $A = (a_{jk})$ over $\mathbf{C}$ such that

$$\sum_{j=1}^{\infty}\sum_{k=1}^{\infty} |a_{jk}|^2 < \infty.$$

We define a scalar product by

$$(A, B) := \mathrm{tr}(AB^*)$$

where tr denotes the trace and B^* denotes the conjugate transpose matrix of B. We recall that $\mathrm{tr}(C + D) = \mathrm{tr}C + \mathrm{tr}D$ where C and D are infinite dimensional matrices. The infinite dimensional unit matrix does not belong to this Hilbert space. ♣

Example 6. Let D be an open set of the Euclidean space $\mathbf{R}^n$. Now $L_2(D)^{pq}$ denotes the space of all $q \times p$ matrix functions Lebesgue measurable on D such that

$$\int_D \mathrm{tr} f(x) f(x)^* dm < \infty$$

where m denotes the Lebesgue measure, * denotes the conjugate transpose, and tr is the trace of the $q \times q$ matrix. We define the scalar product as

$$(f, g) := \int_D \mathrm{tr} f(x) g(x)^* dm.$$

Then $L_2(D)^{pq}$ is a Hilbert space. ♣

Theorem. All complex infinite dimensional Hilbert spaces are isomorphic to $l_2(\mathbf{N})$ and consequently are mutually isomorphic.

Definition. Let S be a subset of the Hilbert space $\mathcal{H}$. The subset S is dense in $\mathcal{H}$ if for every $f \in \mathcal{H}$ there exists a Cauchy sequence $\{f_j\}$ in S such that $f_j \to f$ as $j \to \infty$.

Definition. A Hilbert space is *separable* if it contains a countable dense subset $\{f_1, f_2, \ldots\}$

Example 1. The set of all $\mathbf{u} = (u_1, u_2, \ldots)^T$ in $l_2(\mathbf{N})$ with only finitely many nonzero components u_j is dense in $l_2(\mathbf{N})$. ♣

Example 2. Let $C^1_{(2)}(\mathbf{R})$ be the linear space of the once continuously differentiable functions that vanish at infinity together with their first derivative and are square integrable. Then $C^1_{(2)}(\mathbf{R})$ is dense in $L_2(\mathbf{R})$. ♣

In almost all applications in quantum mechanics the underlying Hilbert space is separable.

Definition. A *subspace* $\mathcal{K}$ of a Hilbert space $\mathcal{H}$ is a subset of vectors which themselves forms a Hilbert space.

It follows from this definition that, if $\mathcal{K}$ is a subspace of $\mathcal{H}$, then so too is the set $\mathcal{K}^\perp$ of vectors orthogonal to all those in $\mathcal{K}$. The subspace $\mathcal{K}^\perp$ is termed the *orthogonal complement* of $\mathcal{K}$ in $\mathcal{H}$. Moreover, any vector f in $\mathcal{H}$ may be uniquely decomposed into components $f_\mathcal{K}$ and $f_{\mathcal{K}^\perp}$, lying in $\mathcal{K}$ and $\mathcal{K}^\perp$, respectively, i.e.

$$f = f_\mathcal{K} + f_{\mathcal{K}^\perp}.$$

Example. Consider the Hilbert space $\mathcal{H} = l_2(\mathbf{N})$. Then the vectors

$$\mathbf{u}^T = (u_1, u_2, \ldots, u_N, 0, \ldots)$$

with $u_n = 0$ for $n > N$, form a subspace $\mathcal{K}$. The orthogonal complement $\mathcal{K}^\perp$ of $\mathcal{K}$ then consists of the vectors

$$(0, \ldots, 0, u_{N+1}, u_{N+2}, \ldots)$$

with $u_n = 0$ for $n \leq N$. ♣

Definition. A sequence $\{\phi_j\}$, $j \in I$ and $\phi_j \in \mathcal{H}$ is called an *orthonormal sequence* if

$$(\phi_j, \phi_k) = \delta_{jk}$$

where I is a countable index set and δ_{jk} denotes the *Kronecker delta*, i.e.

$$\delta_{jk} := \begin{cases} 1 & \text{for} \quad j = k \\ 0 & \text{for} \quad j \neq k \end{cases}$$

Definition. An orthonormal sequence $\{\phi_j\}$ in $\mathcal{H}$ is an *orthonormal basis* if every $f \in \mathcal{H}$ can be expressed as

$$f = \sum_{j \in I} a_j \phi_j \qquad I: \text{ Index set}$$

for some constants $a_j \in \mathbf{C}$. The expansion coefficients a_j are given by

$$a_j := (f, \phi_j)$$

Example 1. Consider the Hilbert space $\mathcal{H} = \mathbf{C}^2$. The scalar product is defined as

$$(\mathbf{u}, \mathbf{v}) := \sum_{j=1}^{2} u_j \bar{v}_j.$$

An orthonormal basis in $\mathcal{H}$ is given by

$$\mathbf{e}_1 = \frac{1}{\sqrt{2}} \begin{pmatrix} 1 \\ i \end{pmatrix}, \qquad \mathbf{e}_2 = \frac{1}{\sqrt{2}} \begin{pmatrix} 1 \\ -i \end{pmatrix}.$$

Let

$$\mathbf{u} = \begin{pmatrix} 1 \\ 2 \end{pmatrix}.$$

Then the expansion coefficients are given by

$$a_1 = (\mathbf{u}, \mathbf{e}_1) = \frac{1}{\sqrt{2}}(1 - 2i), \qquad a_2 = (\mathbf{u}, \mathbf{e}_2) = \frac{1}{\sqrt{2}}(1 + 2i).$$

Consequently

$$\begin{pmatrix} 1 \\ 2 \end{pmatrix} = \frac{1}{\sqrt{2}}(1 - 2i)\mathbf{e}_1 + \frac{1}{\sqrt{2}}(1 + 2i)\mathbf{e}_2. \quad ♣$$

Example 2. Let $\mathcal{H} = L_2(-\pi, \pi)$. Then an orthonormal basis is given by

$$\left\{ \phi_k(x) = \frac{1}{\sqrt{2\pi}} \exp(ikx) \quad : \quad k \in \mathbf{Z} \right\} .$$

Let $f \in L_2(-\pi, \pi)$ with $f(x) = x$. Then

$$a_k = (f, \phi_k) = \int_{-\pi}^{\pi} f \bar{\phi}_k dx = \frac{1}{\sqrt{2\pi}} \int_{-\pi}^{\pi} x \exp(-ikx) dx. \quad \clubsuit$$

Remark. We call the expansion

$$f = \sum_{k \in \mathbf{Z}} (f, \phi_k) \phi_k$$

the *Fourier expansion* of f.

Theorem. Every separable Hilbert space has at least one orthonormal basis.

Inequality of Schwarz. Let $f, g \in \mathcal{H}$. Then

$$|(f, g)| \leq \|f\| \cdot \|g\|$$

Triangle inequality. Let $f, g \in \mathcal{H}$. Then

$$\|f + g\| \leq \|f\| + \|g\| .$$

Let $B = \{ \phi_n \ : \ n \in I \}$ be an orthonormal basis in a Hilbert space $\mathcal{H}$. I is the countable index set. Then

$$(1) \qquad (\phi_n, \phi_m) = \delta_{nm}$$

$$(2) \qquad \bigwedge_{f \in \mathcal{H}} \qquad f = \sum_{n \in I} (f, \phi_n) \phi_n$$

$$(3) \qquad \bigwedge_{f,g \in \mathcal{H}} \qquad (f, g) = \sum_{n \in I} \overline{(f, \phi_n)} (g, \phi_n)$$

$$(4) \qquad \left(\bigwedge_{\phi_n \in B} \quad (f, \phi_n) = 0 \right) \quad \Rightarrow \quad f = 0$$

$$(5) \qquad \bigwedge_{f \in \mathcal{H}} \|f\|^2 = \sum_{n \in I} |(f, \phi_n)|^2$$

Remark. Equation (3) is called *Parseval's relation.*

4.2 Hilbert Tensor Products of Hilbert Spaces

In this section we introduce the concept of the tensor product of Hilbert spaces (Prugovečki [28]). Then we give some applications.

Definition. Let $\mathcal{H}_1, \ldots, \mathcal{H}_n$ be Hilbert spaces with inner products $(\cdot,\cdot)_1, \ldots, (\cdot,\cdot)_n$, respectively. Let $\mathcal{H}_1 \otimes_a \cdots \otimes_a \mathcal{H}_n$ be the algebraic tensor product of $\mathcal{H}_1, \ldots, \mathcal{H}_n$. We denote the inner product in the algebraic tensor product space by $(\cdot,\cdot)$ and define

$$(f,g) := \prod_{k=1}^{n} (f^{(k)}, g^{(k)})_k$$

for f and g of the form

$$f = f^{(1)} \otimes \cdots \otimes f^{(n)}, \qquad g = g^{(1)} \otimes \cdots \otimes g^{(n)}.$$

The equation defines a unique inner product. The *Hilbert tensor product* $\mathcal{H}_1 \otimes \cdots \otimes \mathcal{H}_n$ of the Hilbert spaces $\mathcal{H}_1, \cdots, \mathcal{H}_n$ is the Hilbert space which is the completion of the pre-Hilbert space $\mathcal{E}$ with the inner product.

Consider now two Hilbert spaces $L_2(\Omega_1, \mu_1)$ and $L_2(\Omega_2, \mu_2)$. Denote by $\mu_1 \times \mu_2$ the product of the measures μ_1 and μ_2 on the Borel subsets of $\Omega_1 \times \Omega_2$. If $f \in L_2(\Omega_1, \mu_1)$ and $g \in L_2(\Omega_2, \mu_2)$, then $f(x_1)g(x_2)$ represents an element of $L_2(\Omega_1 \times \Omega_2, \mu_1 \times \mu_2)$, which we denote by $f \cdot g$:

$$(f \cdot g)(x_1, x_2) = f(x_1)g(x_2).$$

Theorem. The linear mapping

$$h \mapsto \hat{h}, \qquad h \in \mathcal{H}_1 \otimes_a \mathcal{H}_2, \qquad \hat{h} \in \mathcal{H}_3$$

$$h = \sum_{k=1}^{n} a_k f_k \otimes g_k, \qquad \hat{h} = \sum_{k=1}^{n} a_k f_k \cdot g_k$$

of the algebraic tensor product of $\mathcal{H}_1 = L_2(\Omega_1, \mu_1)$ and $\mathcal{H}_2 = L_2(\Omega_2, \mu_2)$ into

$$\mathcal{H}_3 = L_2(\Omega_1 \times \Omega_2, \mu_1 \times \mu_2)$$

can be extended uniquely to a unitary transformation of $\mathcal{H}_1 \otimes \mathcal{H}_2$ onto $\mathcal{H}_3$. For the proof we refer to Prugovečki [28].

If $\mathcal{H}_k$ is separable, then there is a countable orthogonal basis

$$\{ \, e_i^{(k)} \, : \, i \in \mathcal{U}_k \, \}$$

in $\mathcal{H}_k$.

Theorem. The Hilbert tensor product $\mathcal{H}_1 \otimes \cdots \otimes \mathcal{H}_n$ of separable Hilbert spaces $\mathcal{H}_1, \ldots, \mathcal{H}_n$ is separable; if $\{ e_i^{(k)} : i \in \mathcal{U}_k \}$ is an orthonormal basis in the $\mathcal{H}_k$, then

$$\{ e_{i_1}^{(1)} \otimes \cdots \otimes e_{i_n}^{(n)} : i_1 \in \mathcal{U}_1, \ldots, i_n \in \mathcal{U}_n \}$$

is an orthonormal basis in $\mathcal{H}_1 \otimes \cdots \otimes \mathcal{H}_n$. Consequently, the set

$$\mathbf{T} := \{ e_{i_1}^{(1)} \otimes \cdots \otimes e_{i_n}^{(n)} : i_1 \in \mathcal{U}_1, \ldots, i_n \in \mathcal{U}_n \}$$

is also countable. In addition, $\mathbf{T}$ is an orthonormal system:

$$(e_{i_1}^{(1)} \otimes \cdots \otimes e_{i_n}^{(n)}, e_{j_1}^{(1)} \otimes \cdots \otimes e_{j_n}^{(n)}) = (e_{i_1}^{(n)}, e_{j_1}^{(1)})_1 \cdots (e_{i_n}^{(n)}, e_{j_n}^{(n)})_n = \delta_{i_1 j_1} \cdots \delta_{i_n j_n}.$$

Let the operator $\hat{A}_k$ be an observable of the system $\mathcal{G}_k$. In the case where $\mathcal{G}_k$ is an independent part of the system $\mathcal{G}$, we can certainly measure $\hat{A}_k$. Hence, this question arises: which mathematical entity related to the Hilbert space $\mathcal{H} = \mathcal{H}_1 \otimes \cdots \otimes \mathcal{H}_n$ represents this observable. To answer this question, we have to define the following concept.

Definition. Let $\hat{A}_1, \ldots, \hat{A}_n$ be n bounded linear operators acting on the Hilbert spaces $\mathcal{H}_1, \ldots, \mathcal{H}_n$, respectively. The tensor product $\hat{A}_1 \otimes \cdots \otimes \hat{A}_n$ of these n operators is that bounded linear operator on $\mathcal{H}_1 \otimes \cdots \otimes \mathcal{H}_n$ which acts on a vector $f_1 \otimes \cdots \otimes f_n$, $f_1 \in \mathcal{H}_1, \ldots, f_n \in \mathcal{H}_n$ in the following way:

$$(\hat{A}_1 \otimes \cdots \otimes \hat{A}_n)(f_1 \otimes \cdots \otimes f_n) = \hat{A}_1 f_1 \otimes \cdots \otimes \hat{A}_n f_n.$$

The above relation determines the operator $\hat{A}_1 \otimes \cdots \otimes \hat{A}_n$ on the set of all vectors of the form $f_1 \otimes \cdots \otimes f_n$, and therefore, due to the presupposed linearity of $\hat{A}_1 \otimes \cdots \otimes \hat{A}_n$, on the linear manifold spanned by all such vectors. Since this linear manifold is dense in $\mathcal{H}_1 \otimes \cdots \otimes \mathcal{H}_n$ and the operator defined above in this manifold is bounded, it has a unique extension to $\mathcal{H}_1 \otimes \cdots \otimes \mathcal{H}_n$. Consequently, the above definition is consistent.

How do we represent in $\mathcal{H} = \mathcal{H}_1 \otimes \cdots \otimes \mathcal{H}_n$ an observable of $\mathcal{S}_k$ representable in $\mathcal{H}_k$ by the bounded operator $\hat{A}_k$? The candidate is

$$\tilde{A}_k := 1 \otimes \cdots \otimes 1 \otimes \hat{A}_k \otimes 1 \otimes \cdots \otimes 1$$

where 1 is the identity operator. We can easily verify that the physically meaningful quantities, i.e., the expectation values are the same for $\hat{A}_k$ when $\mathcal{S}_k$ is an isolated system in the state Ψ_k at t, or when $\mathcal{S}_k$ is at the instant t an independent part of $\mathcal{G}$, which is at t in the state

$$\Psi_1 \otimes \cdots \otimes \Psi_k \otimes \cdots \otimes \Psi_n.$$

Namely, if we take the normalized state vectors $\Psi_1, \ldots, \Psi_n$, then

$$(\Psi_1 \otimes \cdots \otimes \Psi, (1 \otimes \cdots \otimes \hat{A}_k \otimes \cdots \otimes 1)\Psi_1 \otimes \cdots \otimes \Psi_n)$$

$$= (\Psi_1, \Psi_1)_1 \cdots (\Psi_k, A_k \Psi_k)_k \cdots (\Psi_n, \Psi_n)_n = (\Psi_k, \hat{A}_k \Psi_k)_k .$$

Example. Consider the case of n different particles without spin. In that case $\mathcal{H}$ is $L^2(\mathbf{R}^{3n})$, and we can take $\mathcal{H}_1 = \cdots = \mathcal{H}_n = L^2(\mathbf{R}^3)$. The Hilbert space $L_2(\mathbf{R}^{3n})$ is isomorphic to $L^2(\mathbf{R}^3) \otimes \cdots \otimes L_2(\mathbf{R}^3)$ (n factors), and the mapping

$$\psi_1(\mathbf{r}_1) \otimes \cdots \otimes \psi_n(\mathbf{r}_n) \to \psi_1(\mathbf{r}_1) \cdots \psi_n(\mathbf{r}_n)$$

induces a unitary transformation U between these two Hilbert spaces. The Hamilton operator $\hat{H}_k$ is in this case given by the differential operator

$$-\hbar^2/2m_k)\Delta_k$$

and therefore $\hat{H}_1 + \cdots + \hat{H}_n$ is given by

$$-((\hbar^2/2m_1)\Delta_1 + \cdots + (\hbar^2/2m_n)\Delta_n)$$

and represents in this case the kinetic energy of the system $\mathcal{S}$. The interaction term V is determined by the potential $V(\mathbf{r}_1, \ldots, \mathbf{r}_n)$,

$$(V\psi)(\mathbf{r}_1, \ldots, \mathbf{r}_n) = V(\mathbf{r}_1, \ldots, \mathbf{r}_n)\psi(\mathbf{r}_1, \ldots, \mathbf{r}_n).$$

Definition. The closed linear subspace of the Hilbert space

$$\mathcal{H} = \mathcal{H}_1 \otimes \cdots \otimes \mathcal{H}_n, \qquad \mathcal{H}_1 \equiv \cdots \equiv \mathcal{H}_n$$

spanned by all the vectors of the form

$$f_1 \otimes^S \cdots \otimes^S f_n := \frac{1}{\sqrt{n!}} \sum_{(k_1,\ldots,k_n)} f_{k_1} \otimes \cdots \otimes f_{k_n}, \qquad f_1, \ldots, f_n \in \mathcal{H}_1$$

where the sum is taken over all the permutations $(k_1, \ldots, k_n)$ of $(1, \ldots, n)$, is called the *symmetric tensor product* of $\mathcal{H}_1, \ldots, \mathcal{H}_n$, and it is denoted by

$$\mathcal{H}_1 \otimes^S \cdots \otimes^S \mathcal{H}_n \quad \text{or} \quad \mathcal{H}_1^{\otimes^S n}.$$

Similarly, in the case of the set of all vectors in $\mathcal{H}$ of the form

$$f_1 \otimes^A \cdots \otimes^A f_n := \frac{1}{\sqrt{n!}} \sum_{(k_1,\ldots,k_n)} \pi(k_1, \ldots, k_n) f_{k_1} \otimes \cdots \otimes f_{k_n}, \quad f_1, \ldots, f_n \in \mathcal{H}_1$$

where

$$\pi(k_1, \ldots, k_n) := \begin{cases} +1 & \text{if } (k_1, \ldots, k_n) \text{ is even} \\ -1 & \text{if } (k_1, \ldots, k_n) \text{ is odd} \end{cases}$$

the closed linear subspace spanned by this set is called the *antisymmetric tensor product* of $\mathcal{H}_1, \ldots, \mathcal{H}_n$, and it is denoted by

$$\mathcal{H}_1 \otimes^A \cdots \otimes^A \mathcal{H}_n \quad \text{or} \quad \mathcal{H}_1^{\otimes^A n}.$$

The factor $(n!)^{-1/2}$ has been introduced for the sake of convenience in dealing with orthonormal bases. In case that $f_1, \ldots, f_n$ are orthogonal to each other and normalized in

$\mathcal{H}_1$, then due to this factor, the symmetric and antisymmetric tensor products of these vectors will be also normalized.

If we take the inner product

$$(f_1 \otimes^S \cdots \otimes^S f_n, f_1 \otimes^A \cdots \otimes^A f_n)$$

we obtain zero. This implies that $\mathcal{H}_1^{\otimes^S n}$ is orthogonal to $\mathcal{H}_1^{\otimes^A n}$, when these spaces are treated as subspaces of $\mathcal{H}_1^{\otimes n}$.

If $\hat{A}_1, \ldots, \hat{A}_n$ are bounded linear operators on $\mathcal{H}_1$, then $\hat{A}_1 \otimes \cdots \otimes \hat{A}_n$ will not leave, in general, $\mathcal{H}_1^{\otimes^S n}$ or $\mathcal{H}_1^{\otimes^A n}$ invariant. We can define, however, the symmetric tensor product

$$\hat{A}_1 \otimes^S \cdots \otimes^S \hat{A}_n = \sum_{(k_1,\ldots,k_n)} \hat{A}_{k_1} \otimes \cdots \otimes \hat{A}_{k_n}$$

which leaves $\mathcal{H}_1^{\otimes^A n}$ and $\mathcal{H}_1^{\otimes^S n}$ invariant, and which can be considered therefore as defining linear operators on these respective spaces, by restricting the domain of definition of this operator to these spaces.

Example. Consider the Hilbert space $\mathcal{H} = \mathbf{R}^2$ and

$$\begin{pmatrix} a \\ b \end{pmatrix}, \begin{pmatrix} c \\ d \end{pmatrix} \in \mathbf{R}^2.$$

Let

$$\mathbf{u} = \begin{pmatrix} a \\ b \end{pmatrix} \otimes \begin{pmatrix} c \\ d \end{pmatrix} + \begin{pmatrix} c \\ d \end{pmatrix} \otimes \begin{pmatrix} a \\ b \end{pmatrix}$$

and

$$\mathbf{v} = \begin{pmatrix} a \\ b \end{pmatrix} \otimes \begin{pmatrix} c \\ d \end{pmatrix} - \begin{pmatrix} c \\ d \end{pmatrix} \otimes \begin{pmatrix} a \\ b \end{pmatrix}.$$

Then $\mathbf{u}, \mathbf{v} \in \mathbf{R}^4$. We find

$$(\mathbf{u}, \mathbf{v}) = 0$$

since

$$\mathbf{u} = \begin{pmatrix} 2ac \\ ad + bc \\ bc + ad \\ 2bd \end{pmatrix}, \qquad \mathbf{v} = \begin{pmatrix} 0 \\ ad - bc \\ bc - ad \\ 0 \end{pmatrix}. \qquad \clubsuit$$

4.3 Spin and Statistics for the n-Body Problem

As an illustration of the above consideration on the connection between spin and statistics we shall formulate the wave mechanics for n identical particles of spin σ. We mechanics postulates that a state vector will be represented in this case by a wavefunction

$$\psi(\mathbf{r}_1, s_1, \ldots, \mathbf{r}_n, s_n), \quad \mathbf{r}_k \in \mathbf{R}^3, \qquad s_k = -\sigma, -\sigma+1, \ldots, \sigma$$

which is Lebesgue square integrable and symmetric

$$\psi(\ldots, \mathbf{r}_i, s_i, \ldots, \mathbf{r}_j, s_j, \ldots) = -\psi(\ldots, \mathbf{r}_j, s_j, \ldots, \mathbf{r}_i, s_i, \ldots)$$

if σ is half-integer. Thus the Hilbert spaces of functions in which the inner product is taken to be

$$(f,g) = \sum_{s_1=-\sigma}^{+\sigma} \cdots \sum_{s_n=-\sigma}^{+\sigma} \int_{\mathbf{R}^{3n}} f^*(\mathbf{r}_1, s_1, \ldots, \mathbf{r}_n, s_n) g(\mathbf{r}_1, s_1, \ldots, \mathbf{r}_n, s_n) d\mathbf{r}_1 \ldots d\mathbf{r}_n$$

are unitarily equivalent under the unitary transformation induced by the mapping

$$\psi(\mathbf{r}_1, s_1) \otimes \cdots \otimes \psi(\mathbf{r}_n, s_n) \to \psi\mathbf{r}_1, s_1) \cdots \psi(\mathbf{r}_n, s_n)$$

to the spaces $\mathcal{H}_1^{\otimes^S n}$ and $\mathcal{H}_1^{\otimes^A n}$, respectively, where

$$\mathcal{H}_1 = L_2(\mathbf{R}^3) \oplus \cdots \oplus L_2(\mathbf{R}^3), \qquad (2\sigma + 1 \text{ terms}).$$

The Hamilton operator $\hat{H}$ of the system is taken to be of the form

$$\hat{H} = \hat{T} + \hat{V}$$

where $\hat{T}$ is the kinetic energy operator given by

$$(-\hbar^2/2m)(\Delta_1 + \cdots + \Delta_n).$$

Thus, $\hat{T}$ is already symmetric with respect to the n particles. If $\hat{V}$ is the potential energy given by a potential $\hat{V}(\mathbf{r}_1, s_1, \ldots, \mathbf{r}_n, s_n)$, then the principle of indistinguishability of identical particles requires that this function is symmetric under any permutation of the indices $1, \ldots, n$.

Example. We calculate the eigenvalues of the Hamilton operator

$$\hat{H} := \lambda(S_x \otimes \hat{L}_x + S_y \otimes \hat{L}_y + S_z \otimes \hat{L}_z)$$

where we consider a subspace G_1 of the Hilbert space $L_2(S^2)$ with

$$S^2 := \{\, (x,y,z) \; : \; x^2 + y^2 + z^2 = 1 \,\}.$$

Here $\otimes$ denotes the *tensor product.* The linear operators $\hat{L}_x$, $\hat{L}_y$, $\hat{L}_z$ act in the subspace G_1. A basis of subspace G_1 is

$$Y_{1,0} = \sqrt{\frac{3}{4\pi}}\cos\theta, \qquad Y_{1,1} = -\sqrt{\frac{3}{8\pi}}\sin\theta e^{i\phi}, \qquad Y_{1,-1} = \sqrt{\frac{3}{8\pi}}\sin\theta e^{-i\phi}.$$

The operators (matrices) S_x, S_y, S_z act in the Hilbert space $\mathbf{C}^2$ with the basis

$$\begin{pmatrix} 1 \\ 0 \end{pmatrix}, \qquad \begin{pmatrix} 0 \\ 1 \end{pmatrix}.$$

The matrices S_z, S_+ and S_- are given by

$$S_z := \frac{1}{2}\hbar\begin{pmatrix} 1 & 0 \\ 0 & -1 \end{pmatrix}, \qquad S_+ := \hbar\begin{pmatrix} 0 & 1 \\ 0 & 0 \end{pmatrix}, \qquad S_- := \hbar\begin{pmatrix} 0 & 0 \\ 1 & 0 \end{pmatrix}$$

where $S_\pm := S_x \pm iS_y$. The operators $\hat{L}_x$, $\hat{L}_y$ and $\hat{L}_z$ take the form

$$\hat{L}_z := -i\hbar\frac{\partial}{\partial\phi}$$

$$\hat{L}_+ := \hbar e^{i\phi}\left(\frac{\partial}{\partial\theta} + i\,\cot\theta\frac{\partial}{\partial\phi}\right)$$

$$\hat{L}_- := \hbar e^{-i\phi}\left(-\frac{\partial}{\partial\theta} + i\,\cot\theta\frac{\partial}{\partial\phi}\right)$$

where

$$\hat{L}_\pm := \hat{L}_x \pm i\hat{L}_y.$$

We give an interpretation of the Hamilton operator $\hat{H}$ and of the subspace G_1. The Hamilton operator can be written as

$$\hat{H} = \lambda(S_z \otimes \hat{L}_z) + \frac{\lambda}{2}(S_+ \otimes \hat{L}_- + S_- \otimes \hat{L}_+).$$

In the tensor product space $\mathbf{C}^2 \otimes G_1$ a basis is given by

$$|1\rangle = \begin{pmatrix} 1 \\ 0 \end{pmatrix} \otimes Y_{1,0}, \qquad |2\rangle = \begin{pmatrix} 1 \\ 0 \end{pmatrix} \otimes Y_{1,-1}, \qquad |3\rangle = \begin{pmatrix} 1 \\ 0 \end{pmatrix} \otimes Y_{1,1}$$

$$|4\rangle = \begin{pmatrix} 0 \\ 1 \end{pmatrix} \otimes Y_{1,0}, \qquad |5\rangle = \begin{pmatrix} 0 \\ 1 \end{pmatrix} \otimes Y_{1,-1}, \qquad |6\rangle = \begin{pmatrix} 0 \\ 1 \end{pmatrix} \otimes Y_{1,1}.$$

In the following we use

$$\hat{L}_+ Y_{1,1} = 0, \qquad \hat{L}_+ Y_{1,0} = \hbar\sqrt{2}Y_{1,1}, \qquad \hat{L}_+ Y_{1,-1} = \hbar\sqrt{2}Y_{1,0}$$

$$\hat{L}_- Y_{1,1} = \hbar\sqrt{2}Y_{1,0}, \qquad \hat{L}_- Y_{1,0} = \hbar\sqrt{2}Y_{1,-1}, \qquad \hat{L}_- Y_{1,-1} = 0.$$

For the state $|1\rangle$ we find

$$\hat{H}|1\rangle = [\lambda(S_z \otimes \hat{L}_z) + \frac{\lambda}{2}(S_+ \otimes \hat{L}_- + S_- \otimes \hat{L}_+)] \begin{pmatrix} 1 \\ 0 \end{pmatrix} \otimes Y_{1,0}.$$

Thus

$$\hat{H}|1\rangle = \lambda[S_z \begin{pmatrix} 1 \\ 0 \end{pmatrix} \otimes \hat{L}_z Y_{1,0}] + \frac{\lambda}{2}[S_+ \begin{pmatrix} 1 \\ 0 \end{pmatrix} \otimes \hat{L}_- Y_{1,0} + S_- \begin{pmatrix} 1 \\ 0 \end{pmatrix} \otimes \hat{L}_+ Y_{1,0}].$$

Finally

$$\hat{H}|1\rangle = \frac{\lambda}{2} S_- \begin{pmatrix} 1 \\ 0 \end{pmatrix} \otimes \hat{L}_+ Y_{1,0} = \frac{\lambda}{\sqrt{2}}\hbar^2 |6\rangle.$$

Analogously, we find

$$\hat{H}|2\rangle = -\frac{\lambda\hbar^2}{2}|2\rangle + \frac{\lambda\hbar^2}{\sqrt{2}}|4\rangle$$

$$\hat{H}|3\rangle = \frac{\lambda\hbar^2}{2}|3\rangle$$

$$\hat{H}|4\rangle = \frac{\lambda\hbar^2}{2}|2\rangle$$

$$\hat{H}|5\rangle = \frac{\lambda\hbar^2}{2}|5\rangle$$

$$\hat{H}|6\rangle = -\frac{\lambda\hbar^2}{2}|6\rangle + \frac{\lambda\hbar^2}{\sqrt{2}}|1\rangle.$$

Hence the states $|3\rangle$ and $|5\rangle$ are eigenstates with the eigenvalues $E_{1,2} = \lambda\hbar^2/2$. The states $|1\rangle$ and $|6\rangle$ form a two-dimensional subspace. The matrix representation is given by

$$\begin{pmatrix} 0 & \frac{\lambda\hbar^2}{\sqrt{2}} \\ \frac{\lambda\hbar^2}{\sqrt{2}} & -\frac{\lambda\hbar^2}{2} \end{pmatrix}.$$

The eigenvalues are

$$E_{3,4} = -\frac{\lambda\hbar^2}{2} \pm \frac{3\lambda\hbar^2}{4}.$$

Analogously, the states $|2\rangle$ and $|4\rangle$ form a two-dimensional subspace. The matrix representation is given by

$$\begin{pmatrix} -\frac{\lambda\hbar^2}{2} & \frac{\lambda\hbar^2}{\sqrt{2}} \\ \frac{\lambda\hbar^2}{\sqrt{2}} & 0 \end{pmatrix}.$$

The eigenvalues are

$$E_{5,6} = -\frac{\lambda\hbar^2}{2} \pm \frac{3\lambda\hbar^2}{4}.$$

Remark: The Hamilton operator describes the *spin-orbit coupling*. In textbooks we sometimes find the notation $\hat{H} = \lambda \mathbf{S} \cdot \mathbf{L}$. ♣

4.4 Exciton-Phonon Systems

In this section we consider Bose-spin systems which can model exciton-phonon systems. Our first model under investigation is given by

$$\hat{H} = -\Delta I_B \otimes \sigma_x - k(b^\dagger + b) \otimes \sigma_z + \Omega b^\dagger b \otimes I_2$$

where Δ, k, and Ω are constants and $\otimes$ denotes the tensor product. The first term describes transitions between the excitonic sites, the second term represents the exciton-phonon interaction, and the third term is the oscillatory energy. The quantity I_2 is the 2×2 matrix and I_B is the unit operator in the vector space in which the *Bose operators* act. In matrix representation I_B is the infinite unit matrix. The *commutation relations* for the *Bose operators* $b^\dagger$, b are given by

$$[b, b^\dagger] = I_B, \qquad [b, b] = 0, \qquad [b^\dagger, b^\dagger] = 0$$

where $[\,,\,]$ denotes the commutator. For our mathematical treatment it is more convenient to cast the Hamilton operator $\hat{H}$ into a new form. With the help of the unitary transformation

$$\widetilde{H} = \exp(S)\hat{H}\exp(-S), \qquad S = \frac{i\pi}{4} I_B \otimes \sigma_y$$

the Hamilton operator takes the form

$$\hat{H} = -\Delta I_B \otimes \sigma_z + k(b^\dagger + b) \otimes \sigma_x + \Omega b^\dagger b \otimes I_2.$$

Since $\sigma_x \equiv (\sigma_+ + \sigma_-)/2$, we can write

$$\hat{H} = -\Delta I_B \otimes \sigma_z + (k/2)(b^\dagger + b) \otimes (\sigma_+ + \sigma_-) + \Omega b^\dagger b \otimes I_2.$$

Closely related to this Hamilton operator is the Hamilton operator

$$\hat{H} = -\Delta I_B \otimes \sigma_z + (k/2) \otimes (b^\dagger \otimes \sigma_- + b \otimes \sigma_+) + \Omega b^\dagger b \otimes I_2.$$

We now determine the constants of motion for the Hamilton operator $\widetilde{H}$ and $\hat{H}_R$. For the Hamilton operator $\widetilde{H}$ we find $\widetilde{H}$,

$$\hat{P} = \exp(i\pi(b^\dagger b \otimes I_2) + I_B \otimes \sigma_z/2 + I_B \otimes I_2/2)$$

(*parity operator*), and

$$I_B \otimes \sigma^2 \equiv I_B \otimes (\sigma_x^2 + \sigma_y^2 + \sigma_z^2) \equiv I_B \otimes (\sigma_z^2 + (1/2)(\sigma_+\sigma_- + \sigma_-\sigma_+))$$

are constants of motion, i.e.

$$[\widetilde{H}, \widetilde{H}] = 0, \qquad [\widetilde{H}, \hat{P}] = 0, \qquad [\widetilde{H}, I_B \otimes \sigma^2] = 0\,.$$

Notice that $\widetilde{H}$ does not commute with the operator

$$\hat{N} := b^\dagger b \otimes I_2 + I_B \otimes \sigma_z/2\,.$$

The constant of motion $\hat{P}$ enables us to simplify the eigenvalue problem for $\widetilde{H}$. The operator $\hat{P}$ has a discrete spectrum, namely two eigenvalues (infinitely degenerate). With the help of $\hat{P}$ we can decompose our product Hilbert space into two invariant subspaces. In both subspaces we cannot determine the eigenvalues exactly. The Hamilton operator $\hat{H}_R$ commutes with $\hat{H}_R$, $\hat{P}$, $I_B \otimes \sigma^2$, and $\hat{N}$. Notice that

$$[\hat{N}, \hat{P}] = 0 .$$

Owing to the constants of motion $\hat{N}$, we can write the infinite matrix representation of $\hat{H}_R$ as the *direct sum* of 2×2 matrices. Thus we can solve the eigenvalue problem exactly. Next we look at the matrix representation. First we consider the Hamilton operator $\widetilde{H}$. Owing to the constant of motion $\hat{P}$ we can decompose the set of basis vectors

$$\left\{ |n\rangle \otimes \begin{pmatrix} 1 \\ 0 \end{pmatrix}, \ |n\rangle \otimes \begin{pmatrix} 0 \\ 1 \end{pmatrix} \quad n = 0, 1, 2, \ldots \right\}$$

into two subspaces S_1 and S_2, each of them belonging to an $\hat{H}$-invariant subspace

$$S_1 := \left\{ |0\rangle \otimes \begin{pmatrix} 0 \\ 1 \end{pmatrix}, \ |1\rangle \otimes \begin{pmatrix} 1 \\ 0 \end{pmatrix}, \ |2\rangle \otimes \begin{pmatrix} 0 \\ 1 \end{pmatrix}, \ldots \right\}$$

$$S_2 := \left\{ |0\rangle \otimes \begin{pmatrix} 1 \\ 0 \end{pmatrix}, \ |1\rangle \otimes \begin{pmatrix} 0 \\ 1 \end{pmatrix}, \ |2\rangle \otimes \begin{pmatrix} 1 \\ 0 \end{pmatrix}, \ldots \right\}$$

The corresponding infinite symmetric matrices are tridiagonal. For the subspace S_1 we have

$$\widetilde{H}_{n,n} = (-1)^n \Delta + n\Omega, \qquad \widetilde{H}_{n+1,n} = \widetilde{H}_{n,n+1} = k(n+1)^{1/2}$$

where $n = 0, 1, 2, \ldots$. For the subspace S_2 we find the matrix representation

$$\widetilde{H}_{n,n} = -(-1)^n \Delta + n\Omega, \qquad \widetilde{H}_{n+1,n} = \widetilde{H}_{n,n+1} = k(n+1)^{1/2}.$$

Let us consider the Hamilton operator $\hat{H}_R$. For the subspace S_1 we find the matrix representation

$$\begin{pmatrix} -\Delta & k \\ k & \Delta + \Omega \end{pmatrix} \oplus \begin{pmatrix} -\Delta + 2\Omega & \sqrt{3}k \\ k & \Delta + 3\Omega \end{pmatrix} \oplus \cdots \begin{pmatrix} -\Delta + 2n\Omega & \sqrt{2n+1}k \\ \sqrt{2n+1}k & \Delta + (2n+1)\Omega \end{pmatrix} \oplus \cdots$$

where $\oplus$ denotes the direct sum. For the subspace S_2 we obtain

$$\Delta \oplus \begin{pmatrix} -\Delta + \Omega & \sqrt{2}k \\ \sqrt{2}k & \Delta + 2\Omega \end{pmatrix} \oplus \cdots \begin{pmatrix} -\Delta + (2n+1)\Omega & \sqrt{2n+2}k \\ \sqrt{2n+2}k & \Delta + (2n+2)\Omega \end{pmatrix} \oplus \cdots$$

Consequently, we can solve the eigenvalue problem exactly. For example the eigenvalues in the subspace S_1 are given by

$$E_{n\pm} = \left(2n + \frac{1}{2}\right)\Omega \pm \left(\left(\Delta + \frac{\Omega}{2}\right)^2 + (2n+1)k^2\right)^{1/2}$$

where $n = 0, 1, 2, \ldots$.

4.5 Interpretation of Quantum Mechanics

In this section we discuss the interpretation of quantum mechanics and the application of the Kronecker and tensor product. We follow the articles of Elby and Bub [12] and Kent [19].

Quantum mechanics gives us well-understood rules for calculating probabilities, but no precise mathemematical criterion for a unique sample space of events (or physical properties, or propositions) over which a probability distribution is defined. This has motivated many attempts to find mathematical structures which could be used to define a distribution.

Example. Consider a spin-$\frac{1}{2}$ particle initially described by a superposition of eigenstates of S_z, the z component of spin:

$$|\Phi\rangle = c_1|S_z = +\rangle + c_2|S_z = -\rangle.$$

Let $|R = +\rangle$ and $|R = -\rangle$ denote the "up" and "down" pointer-reading eigenstates of an S_z-measuring apparatus. According to quantum mechanics (with no wave-function collapse), if the apparatus ideally measures the particle, the combined system evolves into an entangled superposition,

$$|\varphi\rangle = c_1|S_z = +\rangle \otimes |R = +\rangle + c_2|S_z = -\rangle \otimes |R = -\rangle.$$

Thus after the measurement, the pointer reading should be definite. According to the "orthodox" value-assignment rule, however, the pointer reading is definite only if the quantum state is an eigenstate of $\hat{R}$, the pointer-reading operator. Since $|\varphi\rangle$ is not an eigenstate of $\hat{R}$, the pointer reading is indefinite. The interpretations of quantum mechanics attempt to deal with this aspect of the measurement problem. Their solutions run into a technical difficulty which is called the *basis degeneracy problem.* ♣

One such mathematical structure which could be used to define mathematical structures which could define a distribution follows from Schmidt's observation that, if the Hilbert spae $\mathcal{H}$ is written as the tensor product of spaces $\mathcal{H}_1$ and $\mathcal{H}_2$, a generic state in $\mathcal{H}$ has a unique decomposition into orthonormal basis vectors of a particular type. The Schmidt decomposition has often been used as a technical tool for studying models of system-apparatus measurement. It has been suggested that the Schmidt decomposition contains the complete description of the physical properties of a system state. This is called the *modal interpretation.* However it seems to be no more than a vague and tentative suggestion, since the modalists have so far been unable to produce a natural joint probability distribution for the physical properties of a system at several times, even when these properties all correspond to a single Hilbert space on an equal footing. We assume that a particular splitting is somehow specified. If a set of possible statements about physics can be represented by sets of projective decompositions at various times, and if these projections decohere, then the decoherence functional defines a natural joint probability distribution. Since the Schmidt decompositions are projective, this gives a natural condition and any hypothetical joint probability distribution satisfying the modalists' single-time axiom: when a subset of the Schmidt projections is consistent, the marginal distribution for that subset should be defined by the decoherence functional. This appears to be a rather strong condition, since one can easily find sets of projective decompositions for which it is satisfied by no joint probability distribution. It is natural to ask what the consistency condition implies when the sets of projections are defined by the Schmidt decompositions at each point in time. The answer is that, while the condition does restrict the possible probability joint distribution, it is always possible to find joint distributions satisfying both the consistency condition and the modalist' single-time axiom.

We work in the Heisenberg picture, and consider a closed quantum system with Hilbert space $\mathcal{H}$, in the state ψ, evolving under Hamilton operator $\hat{H}$ from time $t = 0$ onwards. We suppose that an isomorphism

$$\mathcal{H} \simeq \mathcal{H}_1 \otimes \mathcal{H}_2$$

is given at $t = 0$; we write $\dim(\mathcal{H}_j) = n_j$ and suppose that $n_1 \leq n_2$. Let $\{\mathbf{v}_k^j : 1 \leq k \leq n_j\}$ be orthonormal bases of $\mathcal{H}_j$ for $j = 1, 2$, so that $\{\mathbf{v}_k^1 \otimes \mathbf{v}_l^2\}$ forms an orthonormal basis for $\mathcal{H}$. With respect to this isomorphism, the *Schmidt decomposition* of ψ at time t is an expression of the form

$$|\psi(t)\rangle = \sum_{k=1}^{n_1} (p_k(t))^{1/2} \exp(-i\hat{H}t/\hbar)|\mathbf{w}_k^1(t)\rangle \otimes |\mathbf{w}_k^2(t)\rangle$$

where $\{\mathbf{w}_k^1\}$ form an orthonormal basis of $\mathcal{H}_1$ and $\{\mathbf{w}_k^2\}$ form part of an orthonormal basis of $\mathcal{H}_2$. We can take the function $p_k(t)$ to be real and positive, and we take the positive square root. For fixed time t, any decomposition of the form given above then has the same list of weights $\{p_k(t)\}$, and the decomposition is unique provided that this list is non-degenerate. We write

$$|\psi_k(t)\rangle = \exp(-i\hat{H}t/\hbar)|\mathbf{w}_k^1(t)\rangle \otimes |\mathbf{w}_k^2(t)\rangle.$$

Let $W(t)$ be the set of Schmidt weights at time t: that is, the list of $\{p_k(t)\}$ with repetitions deleted. The *Schmidt projections* at time t are defined to be the set of projections $P_p(t)$ onto subspaces of the form span $\{\psi_k(t) : p_k(T) = p\}$, for $p \in W(t)$, together with the projection

$$P_0(t) = (1 - \sum_{p \in W(t)} P_p(t)).$$

Together, these define a projective decomposition of the identity, which we write as

$$\sigma(t) = \{\, P_p(t) : p = 0 \quad \text{or} \quad p \in W(t) \,\}.$$

Kent [19] gives a consistency criterion for the probability distribution on time-dependent Schmidt trajectories and shows that it can be satisfied.

Elby and Bub [12] showed that when a quantum state vector can be written in the *triorthogonal form*

$$|\Psi\rangle = \sum_i c_i |\mathbf{u}_i\rangle \otimes |\mathbf{v}_i\rangle \otimes |\mathbf{w}_i\rangle$$

then there exists no other triorthogonal basis in terms of which $|\Psi\rangle$ can be expanded. Several interpretations of quantum mechanics can make use of this special basis. Many-world adherents can claim that a branching of worlds occurs in the preferred basis picked out by the unique triorthogonal decomposition. Modal interpreters can postulate that the triorthogonal basis helps to pick out which observables possess definite values at a given time. And decoherence theorists can cite the uniqueness of the triorthogonal decomposition as a principled reason for asserting that pointer readings become "classical" upon interacting with the environment. Many-world, decoherence, and modal interpretations of quantum mechanics suffer from a basis degeneracy problem arising from the nonuniqueness of some biorthogonal decompositions.

The basis degeneracy problem arises in the context of many-world interpretations as follows. Many-world interpretations address the measurement problem by hypothesizing that when the combined system occupies state $|\varphi\rangle$, the two branches of the superposition split into separate worlds, in some sense. The pointer reading becomes definite relative to its branch. For instance, in the "up" world, the particle has spin up and the apparatus possesses the corresponding pointer reading. In this way, many-world interpreters explain why we always see definite pointer readings, instead of superpositions. This approach suffers from a well-known technical problem, the basis degeneracy problem, which arises from the nonuniqueness of some biorthogonal decompositions. According to the biorthogonal decomposition theorem, any quantum state vector describing two systems can, for a certain choice of bases, be expanded in the simple form

$$\sum_i c_i |\mathbf{u}_i\rangle \otimes |\mathbf{v}_i\rangle$$

where the $\{\, |\mathbf{u}_i\rangle \,\}$ and $\{\, |\mathbf{v}_i\rangle \,\}$ vectors are orthonormal, and are therefore eigenstates of Hermitian operators (observables) $\hat{A}$ and $\hat{B}$ associated with systems 1 and 2, respectively. This biorthogonal expansion picks out the Schmidt basis. The basis degeneracy problem arises because the biorthogonal decomposition is unique just in case all of the nonzero $|c_1|$'s are different. When $|c_1| = |c_2|$, we can biorthogonally expand φ in an infinite number of bases. For instance, we can construct $\hat{S}_x$ eigenstates out of linear combinations of $\hat{S}_z$ eigenstates. And, similarly, we can introduce a new apparatus observable $\hat{R}'$, whose eigenstates are superpositions of pointer-reading eigenstates:

$$|S_x = \pm\rangle = \frac{1}{\sqrt{2}}(|S_z = +\rangle \pm |S_z = -\rangle), \qquad |R' = \pm\rangle = \frac{1}{\sqrt{2}}(|R = +\rangle \pm |R = -\rangle).$$

When $c_1 = c_2 = 1/\sqrt{2}$, we can obtain

$$|\varphi\rangle = \frac{1}{\sqrt{2}}(|S_z = +\rangle \otimes |R' = +\rangle + |S_x = -\rangle \otimes |R' = -\rangle).$$

These two ways of writing $|\varphi\rangle$ correspond to two ways of writing the reduced density operator ρ_a that describes the apparatus. Taking the trace over the particle's states we obtain the density

$$\rho_a = \frac{1}{2}(|R' = +\rangle\langle R' = +| + R' = -\rangle\langle R' = -|)$$

where the index a indicates the apparatus. We can see that nothing is special about the pointer-reading basis. The formalism gives us no more reason to assert that the universe splits into pointer-reading eigenstates that then it gives us to assert that the universe splits into $\hat{R}'$ eigenstates. For this reason, the basis degeneracy problem leaves many-world interpreters without a purely formal algorithm for deciding how splitting occurs. This solves the basis degeneracy problem for the many-world interpretation. As the decoherence theorists show, when the environment interacts with the combined particle-apparatus system, the following state results

$$|\Psi\rangle = c_1|S_z = +\rangle \otimes |R = +\rangle \otimes |E_+\rangle + c_2|S_z = -\rangle \otimes |R = -\rangle \otimes |E_-\rangle$$

where $|E_\pm\rangle$ is the state of the rest of the universe after the environment interacts with the apparatus. As time passes, these environmental states quickly approach orthogonality: $\langle E_+|E_-\rangle \to 0$. In this limit, we have a triorthogonal decomposition of $|\Psi\rangle$. Even if $c_1 = c_2$, the triorthogonal decomposition is unique. In other words, no transformed bases exist such that $|\Psi\rangle$ can be expanded as

$$d_1|S' = +\rangle \otimes |R' = +\rangle \otimes |E'_+\rangle + d_2|S' = -\rangle \otimes |R' = -\rangle \otimes |E'_-\rangle.$$

Therefore, a "preferred" basis is chosen. Many-world interpreters can postulate that this basis determines the branches into which the universe splits.

Decoherence theorists have been stressing the idea that the environment picks out the pointer-reading basis. An existential interpretation, a sophisticated variant of the many-world view, relies on the environment to select the "correct" basis. This interpretation suffers from a version of the basis degeneracy problem. If $|\Psi\rangle$ describes the universe, and if $\langle E_+|E_-\rangle = 0$, then the reduced density operator $\rho_{p\&a}$ describing the particle and apparatus (found by tracing over the environmental degrees of freedom) is

$$\rho_{p\&a} = |c_1|^2|S_z = +\rangle\langle S_z = +||R = +\rangle\langle R = +| + |c_2|^2|S_z = -\rangle\langle S_z = -||R = -\rangle\langle R = -|$$

the same mixture as would be obtained upon wave-function collapse. If $c_1 = c_2$, however, then we can decompose this mixture into another basis, in which case the pointer reading loses its "special" status. For example, define

$$|q_\pm\rangle := \frac{1}{\sqrt{2}}(|S_z = +\rangle \otimes |R = +\rangle \pm |S_z = -\rangle \otimes |R = -\rangle).$$

If $c_1 = c_2 = 1/\sqrt{2}$, then we can write $\rho_{p\&a}$ as

$$\rho_{p\&a} = \frac{1}{2}(|q_+\rangle\langle q_+| + |q_-\rangle\langle q_-|).$$

Although decoherence-based interpretations can deal with their basis degeneracy problem in many ways, a particularly "clean," formal solution is to invoke the uniqueness of the triorthogonal decomposition of $|\Psi\rangle$. Uniqueness holds even when $c_1 = c_2$.

A third kind of interpretation is modal interpretation that relies on the biorthogonal decomposition theorem. According to most modal interpretations, if

$$\sum_i c_i |\mathbf{u}_i\rangle \otimes |\mathbf{v}_i\rangle$$

is the unique biorthogonal decomposition of the quantum state, then system 1 has a definite value for observable $\hat{A}$, and system 2 has a definite value for for $\hat{B}$. For instance, consider our example given above. According to modal interpretations, if $|c_1| \neq |c_2|$, then the particle has a definite z component of spin, and the apparatus has a definite pointer reading. These possessed values result not for a world splitting; the entangled wave function still exists entirely in our world, and continues to determine the dynamical evolution of the system. Rather, these modally possessed values are a kind of hidden variable. According to modal interpretations, an observable can possess a definite value even when the quantum state is not an eigenstate of that observable. The unique biorthogonal decomposition determines which observables take on definite values. Thus modal interpretations suffer from the basis degeneracy problem, just as many-world interpretations do, when $|c_1| = |c_2|$. When the particle-apparatus system interacts with its environment, it evolves into $|\Psi\rangle$, which is (uniquely) triorthogonally decomposed. By allowing unique triorthogonal decompositions – as well as unique biorthogonal decompositions – to pick out which observables receive definite values, modal interpreters can explain why all ideal measurements have definite results (i.e., after the measurement, the pointer reading is definite). The basis selected by a triorthogonal decomposition never conflicts with the basis picked out by the unique biorthogonal decomposition, when one exists. By proving the uniqueness of the triorthogonal decomposition, we can help many-world, modal, and decoherence interpretations deal with the basis degeneracy problem. Elby and Bub [12] proved that when a quantum state can be written in the triorthogonal form

$$|\Psi\rangle = \sum_i c_i |\mathbf{u}_i\rangle \otimes |\mathbf{v}_i\rangle \otimes |\mathbf{w}_i\rangle$$

then, even if some of the c_i's are equal, no alternative bases exist such that $|\Psi\rangle$ can be rewritten

$$\sum_i d_i |\mathbf{u}_i'\rangle \otimes |\mathbf{v}_i'\rangle \otimes |\mathbf{w}_i'\rangle.$$

Therefore the triorthogonal decomposition picks out a "special" basis. They use this preferred basis to address the basis degeneracy problem.

Chapter 5

C++ Software Implementation

5.1 Abstract Data Type and C++

C++ not only corrects most of the deficiencies found in C, it also introduces many completely new features that were designed for the language to provide data abstraction and object-oriented programming. Here are some of the prominent new features:

- *Classes*, the basic language construct that consists of data structure and operations applicable to the class. A class is an abstract data type.
- *Member variables*, which describe the attributes of the class.
- *Member functions*, which define the permissible operations of the class.
- *Operator overloading*, which gives additional meaning to operators so that they can be used with user-defined data types.
- *Function overloading*, which is similar to operator overloading. It allows one function to have several definitions thereby reducing the need for unusual function names, making code easier to read.
- *Programmer-controlled automatic type conversion*, which allows us to blend user-defined types with the fundamental data types provided by C++.
- *Derived classes*, also known as subclasses, inherit member variables and member functions from their base classes (also known as superclasses). They can be differentiated from their base classes by adding new member variables, member functions or overriding existing functions.
- *Virtual functions*, which allow a derived class to redefine member functions inherited from a base class. Through *dynamic binding*, the run-time system will choose an appropriate function for the particular class.

We implement vectors and matrices as abstract data types in C++. A comprehensive survey on C++ is given by Steeb and Solms [35].

The classes included in the header files

`vector.h`

and

`matrix.h`

have been tested using:

- Edison Design Group C/C++ Front End (version 2.19) on Unix systems.
- Borland C++ (version 4.5) for Windows.
- Borland C++ (version 2.0) for OS/2.
- VisualAge C++ (version 3.0) for OS/2.

The vector and matrix norms are included in the header files

`VecNorm.h`

`MatNorm.h`

5.2 The Vector Class

5.2.1 Abstraction

A vector is a common mathematical structure in linear algebra and vector analysis. This structure could be constructed using arrays. However, C and C++ arrays have some weaknesses. They are effectively treated as pointers. It is therefore useful to introduce a `Vector` class as an abstract data type. With bound checking and mathematical operators overloaded (for example vector addition), we build a comprehensive and type-safe `Vector` class. On the other hand, it could replace the array supported by C and C++ as a collection of data objects.

The `Vector` class is a structure that possesses many interesting properties for manipulation. The behaviours of the `Vector` ADT are summarized as follows:

- It is best implemented as a template class, because it is a *container class* whereby the data items could be of any type.
- The construction of a `Vector` is simple.
- Arithmetic operators such as `+`, `-`, `*`, `/` with `Vector` and numeric constants are available.
- The assignment and modification forms of assignment `=`, `+=`, `-=`, `*=`, `/=` are overloaded. We could also copy one `Vector` to another by using the assignment operator.
- The subscript operator `[]` is overloaded. This is useful for accessing individual data items in the `Vector`.
- The equality (`==`) and inequality (`!=`) operators check if the two vectors contain the same elements in the same order.
- Operations such as scalar product and cross product for `Vector` are available.
- The member function `length()` returns the size of a `Vector` while `resize()` reallocates the `Vector` to the size specified.
- The `Matrix` class is declared as a `friend` of the class. This indicates that the `Matrix` class is allowed to access the `private` region of the class.
- Input and output stream operations with `Vector` are supported.
- The auxiliary file `VecNorm.h` contains different types of norm operators: $||\mathbf{v}||_1$, $||\mathbf{v}||_2$, $||\mathbf{v}||_\infty$ and the normalization function for the `Vector`.

5.2.2 Templates

A *container class* implements some data structures that "contain" other object. Examples of containers include arrays, list, sets and vectors etc. Templates work especially well for containers since the logic to manage a container is often largely independent of its contents. In this section, we see how templates can be used to build one of the fundamental data structures in mathematics — The `Vector` class.

The container we implemented here is *homogeneous*, i.e. it contains objects of just one type as opposed to a container that contains objects of a variety of types. It also has a *value semantics*. Therefore it contains the object itself rather than the reference to the object.

5.2.3 Data Fields

There are only two data fields in the class:

- The variable `size` stores the length of the `Vector`.
- The variable `data` is a pointer to template type `T` that is used to store the data items of the `Vector`. The memory is allocated dynamically so that it fits into the need of the application. For data types that require huge amounts of memory, it is advisable to release the memory as soon as it is no longer in use. This is not possible with static memory allocation. Therefore, most array based data types such as vectors and matrices should use dynamic memory for their data storage.

Note that there is not any item in the data fields that records the lower or upper index bound of the vector, this means that the index will run from $0, 1, \ldots,$`size-1`. To make a vector that starts from an index other than zero, we may introduce a derived class that inherited all the properties and behaviours of the `Vector` class and add an extra data field that indicates the lower index bound of the `Vector`. It is therefore of the utmost importance to declare the data fields as `protected` rather than `private`. This allows the derived classes to access the data fields. The implementation of such a bound vector is left as an exercise for the readers.

5.2.4 Constructors

Whenever an array of n vectors are declared, the compiler automatically invoke the default constructor. Therefore, in order to ensure the proper execution of the class, we need to initialize the data items properly in the default constructor. This includes assigning `data` to `NULL`. This step is crucial or else we may run into some run-time problem.

A common error is to assign a `Vector` to an uninitialized one using the assignment operator `=`, which first frees the old contents (`data`) of the left hand side. But as there is no "old" value, some random value in `data` is freed, probably with disastrous effect. The remedy is to nullify the variable `data` because deleting a `NULL` pointer is perfectly valid and has no side effect.

There are, in fact, two more overloaded constructors:

`Vector(int n)` and `Vector(int n, T value)`

The first constructor allocates `n` memory locations for the `Vector`, whereas the other initializes the data items to `value` on top of that.

The copy constructor `Vector(const Vector& source)` constructs a new `Vector` identical to `source`. It will be invoked automatically to make temporary copies when needed, for example the passing function parameters and return values. It could also be used explicitly during the construction of a `Vector`.

In the following, we listed some common ways to construct a `Vector`:

```
// declare a vector of 10 numbers of type int
Vector<int> u(10);

// declare a vector of 10 numbers and initialize them to 0
Vector<int> v(10,0);

// use a copy constructor to create and duplicate a vector
Vector<int> w(v);
```

Whenever a local `Vector` object passes out of scope, the destructor comes into the play. It releases the array storage in free memory. Otherwise, it will constitute unusable storage, because it is allocated but no pointer points to it.

5.2.5 Operators

Most of the operators applicable to `Vector` are implemented in the class, namely

```
(unary)+, (unary)-, +, -, *, /,
=, +=, -=, *=, /=, ==, !=, |, %, [].
```

Suppose `u`, `v`, `w` are vectors and `c` is a numeric constant, then the available operations are defined as follows:

- `u+v`, `u-v`, `u*v`, `u/v` adds, subtracts, multiplies, or divides corresponding elements of `u` and `v`.
- `u+=v`, `u-=v`, `u*=v`, `u/=v` adds, subtracts, multiplies, or divides corresponding elements of `v` into `u`.
- `u+=c`, `u-=c`, `u*=c`, `u/=c` adds, subtracts, multiplies, or divides each element of `u` with the scalar.

- The assignment operator = should be overloaded in the class. Should one omit to define an assignment operator, the C++ compiler will write one. However, one should bear in mind that the code produced by the compiler simply makes a byte-for-byte copy of the data members. In the case where the class allocates memory dynamically, we usually have to write our own assignment operator. This is because the byte-for-byte operation copies only the memory address of the pointer, not the memory content. It is dangerous to have multiple pointers pointing at the same memory location without a proper management. The same argument applies to the copy constructor.

  Two forms of the assignment operator have been overloaded:

  - `u=v` makes a duplication of `v` into `u`.
  - `u=c` assigns the constant `c` to every entry of `u`.

  Note that the assignment operator is defined such that it returns a reference to the object iteself, thereby allowing constructs like `u = v = w`.

- `u==v` returns *true* if `u` and `v` contain the same elements in the same order and returns *false* otherwise.

- `u!=v` is just the converse of `u==v`.

- We use the symbol `|` as the scalar product operator (also called the inner product). It is defined as $\mathtt{u|v} = \mathtt{u} \cdot \mathtt{v} = \sum_{j=0}^{n-1} u_j v_j$.

- The vector product (also called the cross product) is operated by `%` in the class.

- The `[]` operator allows `u[i]` to access the $\mathtt{i}^{\text{th}}$ entry of the `Vector u`. It must return a reference to, not the value of, the entry because it must be usable on the left-hand side of an assignment. In C++ terminology, it must be an `lvalue`.

The following is an example of the usage of the dot product and cross product of the `Vector` class. Suppose A, B, C, D are four vectors in $\mathbf{R}^3$, then

$$A \times (B \times C) + B \times (C \times A) + C \times (A \times B) = \mathbf{0}$$

$$\begin{aligned}
(A \times B) \times (C \times D) &= B(A \cdot C \times D) - A(B \cdot C \times D) \\
&= C(A \cdot B \times D) - D(A \cdot B \times C) \\
A \cdot (B \times C) &= (A \times B) \cdot C
\end{aligned}$$

The following excerpt program demonstrates that the identities are obeyed for some randomly selected vectors:

```
// vprod.cxx

#include <iostream.h>
#include "Vector.h"

void main()
{
   Vector<double> A(3), B(3), C(3), D(3);

   A[0] = 1.2; A[1] = 1.3; A[2] = 3.4;
   B[0] = 4.3; B[1] = 4.3; B[2] = 5.5;
   C[0] = 6.5; C[1] = 2.6; C[2] = 9.3;
   D[0] = 1.1; D[1] = 7.6; D[2] = 1.8;

   cout << A%(B%C) + B%(C%A) + C%(A%B) << endl;

   cout << (A%B)%(C%D) << endl;
   cout << B*(A|C%D)-A*(B|C%D) << endl;
   cout << C*(A|B%D)-D*(A|B%C) << endl;

   // precedence of | is lower than <<
   cout << (A|B%C) - (A%B|C) << endl;
}
```

```
Result
======
[1.42109e-14]
[0]
[0]

[372.619]
[376.034]
[540.301]

[372.619]
[376.034]
[540.301]

[372.619]
[376.034]
[540.301]

0
```

The small, non-zero value `1.42109e-14` is due to rounding errors. Thus to obtain the correct result, namely the zero vector, we could use the data type `Vector<Rational<int> >`.

5.2.6 Member Functions and Norms

Other than the arithmetic operators, there exist some useful operations for the `Vector` class. Their definitions and properties are listed as follows:

- The function `length()` returns the size of the `Vector`.
- `resize(int n)` sets the `Vector`'s length to `n`. All elements are unchanged, except that if the new size is smaller than the original, then trailing elements are deleted, and if greater, trailing elements are uninitialized.
- `resize(int n, T value)` behaves similar to the previous function except when the new size is greater than the original, trailing elements are initialized to `value`.

In the auxiliary file `VecNorm.h`, we implement three different vector norms and the normalization function:

- `norm1(x)` is defined as $||\mathbf{x}||_1 := |x_1| + |x_2| + \ldots + |x_n|$.
- `norm2(x)` is defined as $||\mathbf{x}||_2 := \sqrt{x_1^2 + x_2^2 + \ldots + x_n^2}$, the return type of `norm2()` is `double`.
- `normI(x)` is defined as $||\mathbf{x}||_\infty := \max\{\,|x_1|,\ |x_2|,\ \ldots,\ |x_n|\,\}$.
- The function `normalize(x)` is used to normalize a vector `x`. The normalized form of the vector `x` is defined as `x/|x|` where `|x|` is the 2-norm of `x`.

In order to have a better understanding about these functions, let us consider some examples:

```
// vnorm.cxx

#include <iostream.h>
#include "Vector.h"
#include "VecNorm.h"

void main()
{
   Vector<int> v;
   v.resize(5,2);
   cout << "The size of vector v is " << v.length() << endl;
   cout << endl;

   Vector<double> a(4,-3.1), b;
   b.resize(4);
   b[0] = 2.3; b[1] = -3.6; b[2] = -1.2; b[3] = -5.5;

   // Different vector norms
   cout << "norm1() of a = " << norm1(a) << endl;
   cout << "norm2() of a = " << norm2(a) << endl;
```

```
    cout << "normI() of a = " << normI(a) << endl;
    cout << endl;

    cout << "norm1() of b = " << norm1(b) << endl;
    cout << "norm2() of b = " << norm2(b) << endl;
    cout << "normI() of b = " << normI(b) << endl;
    cout << endl;

    // The norm2() of normalized vectors a and b is 1
    cout << "norm2() of normalized a = " << norm2(normalize(a)) << endl;
    cout << "norm2() of normalized b = " << norm2(normalize(b)) << endl;
 }

Result
======
The size of vector v is 5

norm1() of a = 12.4
norm2() of a = 6.2
normI() of a = 3.1

norm1() of b = 12.6
norm2() of b = 7.06682
normI() of b = 5.5

norm2() of normalized a = 1
norm2() of normalized b = 1
```

5.2.7 Streams

The overloaded output stream operator `<<` simply exports all the entries in the vector **v** and put them in between a pair of square brackets $[v_0, v_1, v_2, \ldots, v_{n-1}]$.

The input stream operator `>>` first reads in the size of the `Vector` followed by the data entries.

5.3 The Matrix Class

5.3.1 Abstraction

Matrices are two-dimensional arrays with a certain number of rows and columns. They are important structures in linear algebra. To build a matrix class, we do not have to start from scratch. We make use of the advantages (reusability and extensibility) of object-oriented programming to build the new class based on the `Vector` class. Vector is a special case of a matrix with number of column being equal to one. Thus we are able to define a matrix as a vector of vectors.

```
template <class T> class Matrix
{
private:
// Data Fileds
int rowNum, colNum;
Vector<T>* mat;
...
}
```

Notice that we have declared the `Matrix` as a template class. Defining the matrix as a vector of vectors has the advantage that the matrix class methods can use many of the vector operations defined for the vector class. For example, we could perform the vector additions on each row of the matrix to obtain the result of a matrix addition. This greatly reduces the amount of code duplication.

To use the `matrix` class effectively, the users need to familiarize themselves with the behaviours and interfaces of the class. Below, we summarize the properties of the `Matrix` ADT:

- It is implemented as a template class. This indicates that the data type of the data items could be of any type including built-in types and user-defined types.
- There are several simple ways to construct a `Matrix`.
- Arithmetic operators such as `+`, `-`, `*` with `Matrix` and `+`, `-`, `*`, `/` with numeric constants are available.
- The assignment and modification forms of assignment `=`, `+=`, `-=`, `*=` and `/=` are overloaded.
- The Vectorize operator is available.
- The *Kronecker product* of two matrices is supported.
- The subscript operator `[]` is overloaded to access the row vector of the matrix while the parenthesis operator `()` is overloaded to access the column vector of the matrix.

- The equality (==) and inequality (!=) operators check if two matrices are identical.
- The transpose, trace and determinant of a matrix are implemented.
- The *LU decomposition* and inverse of a matrix are implemented as well.
- The member function `resize()` reallocates the memory for row and column vectors according to the new specification provided in the arguments of the function.
- The member functions `rows()` and `cols()` return the number of the row or column of the matrix respectively.
- Input (>>) and output (<<) stream operators are supported.
- The auxiliary file `MatNorm.h` contains the three different matrix norms: $||A||_1$, $||A||_\infty$ and $||A||_H$.

5.3.2 Data Fields

The data fields `rowNum` and `colNum` specify the number of row and column of the matrix respectively.

`Vector<T> *mat` stores the data item of the matrix. It is declared as a vector of vectors. To allocate the memory space for an $m \times n$ matrix, we have to first allocate a vector of m pointers to `Vector`. Then for each pointer, we allocate again a vector of size n. This would result in a total of $m \times n$ memory space for the matrix. After the initialization has been done properly, the matrix is then operational.

5.3.3 Constructors

There are a couple of ways to construct a `Matrix` in the class. However, one prime criterion for a matrix to exist is the specification of the number of its rows and columns:

- `Matrix()` declares a matrix with no size specified. Such a matrix is not usable. To activate the matrix, we make use of a member function called `resize()`, which reallocates the matrix with the number of rows and columns specified in the arguments of the function.
- `Matrix(int nr, int nc)` declares an `nr*nc` matrix with the entries value undefined.
- `Matrix(int nr, int nc, T value)` declares an `nr*nc` matrix with all the entries initialized to `value`.
- `Matrix(const Vector<T> &v)` constructs a matrix from a vector `v`. It is understood that the resultant matrix will contain only one column.
- The copy constructor makes a duplication of a matrix. It is invoked automatically by the compiler when needed and it can be invoked by the user explicitly as well.

- The destructor simply releases the unused memory back to the free memory pool.

Below are some examples on how to declare a `Matrix`:

```
// declare a 2-by-3 matrix of type int
Matrix<int> m(2,3)

// declare a 3-by-4 matrix and initialize the entries to 5
Matrix<int> n(3,4,5)

// duplicate a matrix using the copy constructor
Matrix<int> p(n);

// construct a matrix q from a vector v
Vector<double> v(3,0);
Matrix<double> q(v);
```

5.3.4 Operators

There are many matrix operators implemented in the class, namely

```
(unary)+, (unary)-, +, -, *, /,
=, +=, -=, *=, /=, [], (), ==, !=.
```

Some of the operators are overloaded with more than one meaning! The users are advised to read the documentation carefully before using the class.

In the following, we briefly discuss the behaviours and usages of the operators: Suppose `A`, `B` are matrices, `v` is a vector and `c` is a numeric constant,

- The operations `A+B`, `A-B`, `A*B` adds, subtracts, multiplies two matrices according to their normal definitions.

- The operations `A+c`, `A-c`, `A*c` and `A/c` are defined as `A+cI`, `A-cI`, `A*cI` and `A/cI` respectively where `I` is the identity matrix.

- The operations `c+A`, `c-A`, `c*A` and `c/A` have the similar definitions as above.

- `A=B` makes a duplication of `B` into `A` whereas `A=c` assigns the value `c` to all the entries of the matrix `A`.

- `A+=B`, `A-=B`, `A*=B` are just the modification forms of assignments which perform two operations in one shot. For example, `A+=B` is equivalent to `A = A+B`.

- `A+=c`, `A-=c`, `A*=c`, `A/=c` are just the modification forms of assignments. For example, `A+=c` is equivalent to `A = A+cI`.

- The function `vec(A)` is used to create a vector that contains elements of the matrix `A`, one column after the other (see chapter 1).

- The Kronecker product of two matrices is described in chapter 2. The function `kron(A,B)` is used for calculating the Kronecker product of the matrices A and B.
- The subscript operator `[]` is overloaded to access a specific row vector of a matrix. For example, `A[i]` returns the $\mathtt{i}^{\text{th}}$ row vector of matrix `A`.
- The parenthesis operator `()` is overloaded to access a specific column vector of a matrix. For example, `B(j)` returns the $\mathtt{j}^{\text{th}}$ column vector of matrix `B`.
- The equality (`==`) and inequality (`!=`) operator compare whether the individual entries of two matrices match each other in the right order.

Note that the precedence of `==` and `!=` are lower than the output stream operator `<<`. This means that a pair of brackets is required when the users write statements that resemble the following:

```
cout << (u != v) << endl;
cout << (u == v) << endl;
```

otherwise, the compiler may complain about it.

5.3.5 Member Functions and Norms

Many useful operations have been included in the class. Their properties are described as follows:

- The `transpose()` of an $m \times n$ matrix A is the $n \times m$ matrix A^T such that the ij entry of A^T is the ji entry of A.
- The `trace()` of an $n \times n$ matrix A is the sum of all the diagonal entries of A.
- `determinant()`: The method employed depends on whether the matrix is symbolic or numeric. Since our system is meant to solve also symbolic expressions, we use the Leverrier's method for solving the determinant.
- `LU()` is used to decompose a matrix A into a product of two matrices,

  $$LU = A$$

  where L is a lower triangular and U is an upper triangular matrix. The purpose of such decomposition is to facilitate the process of solving linear algebraic equations shown as follows:

  $$A\mathbf{x} = (LU)\mathbf{x} = L(U\mathbf{x}) = \mathbf{b}$$

  and by first solving for the vector $\mathbf{y}$ with

  $$L\mathbf{y} = \mathbf{b}$$

and then solving

$$U\mathbf{x} = \mathbf{y}.$$

The two linear systems above could be easily solved by forward and backward substitutions.

- The `inverse()` is used to obtain the inverse of an invertible matrix. The methods used for finding the inverse of a matrix are different for the numeric and symbolic matrices. For a numeric matrix, we use the LU decomposition and backward substitution routines, whereas for a symbolic matrix, we use the Leverrier's method.

- `resize()` reallocate the number of row and column according to the new specification provided in the arguments of the function.

- `rows()` and `cols()` return the number of rows and columns of the matrix, respectively.

In the auxiliary file `MatNorm.h`, we implement three different matrix norms:

- `norm1(A)` is defined as the maximum value of the sum of the entries in column vectors,

$$||A||_1 := \max_{1\leq j\leq n}\left(\sum_{i=1}^{n}|a_{ij}|\right)$$

- `normI(A)` is defined as the maximum value of the sum of the entries in row vectors,

$$||A||_\infty := \max_{1\leq i\leq n}\left(\sum_{j=1}^{n}|a_{ij}|\right)$$

- `normH(A)` is the *Hilbert-Schmidt norm.* It is defined as

$$||A||_H := (\mathrm{tr}(A^*A))^{1/2} = (\mathrm{tr}(AA^*))^{1/2} = \sqrt{\sum_{i=1}^{n}\sum_{j=1}^{m}|a_{ij}|^2}$$

Example 1. We demonstrate the usage of the Kronecker products of matrices.

```
// kron.cxx

#include <iostream.h>
#include "Matrix.h"

void main()
{
   Matrix<int> A(2,3), B(3,2), C(3,1), D(2,2);

   A[0][0] = 2; A[0][1] = -4; A[0][2] = -3;
   A[1][0] = 4; A[1][1] = -1; A[1][2] = -2;

   B[0][0] = 2; B[0][1] = -4;
   B[1][0] = 2; B[1][1] = -3;
   B[2][0] = 3; B[2][1] = -1;

   C[0][0] =  2;
   C[1][0] =  1;
   C[2][0] = -2;

   D[0][0] = 2; D[0][1] =  1;
   D[1][0] = 3; D[1][1] = -1;

   cout << kron(A,B) << endl;
   cout << kron(B,A) << endl;
   cout << kron(A,B)*kron(C,D) - kron(A*C,B*D) << endl;
}
```

```
Result
======
[4 -8 -8 16 -6 12]
[4 -6 -8 12 -6 9]
[6 -2 -12 4 -9 3]
[8 -16 -2 4 -4 8]
[8 -12 -2 3 -4 6]
[12 -4 -3 1 -6 2]

[4 -8 -6 -8 16 12]
[8 -2 -4 -16 4 8]
[4 -8 -6 -6 12 9]
[8 -2 -4 -12 3 6]
[6 -12 -9 -2 4 3]
[12 -3 -6 -4 1 2]
```

The last output is the zero matrix.

Example 2. We demonstrate that

$$\mathrm{tr}(AB) = \mathrm{tr}(BA) \qquad \text{and} \qquad \mathrm{tr}(AB) \neq \mathrm{tr}(A)\mathrm{tr}(B)$$

in general.

```
// trace.cxx

#include <iostream.h>
#include "Matrix.h"

void main()
{
   Matrix<int> A(3,3), B(3,3,-1);

   A[0][0] = 2; A[0][1] = -1; A[0][2] =  1;
   A[1][0] = 1; A[1][1] = -2; A[1][2] = -1;
   A[2][0] = 3; A[2][1] =  2; A[2][2] = 2;

   cout << "A =\n" << A << endl;
   cout << "B =\n" << B << endl;
   cout << "tr(A) = " << A.trace() << endl;
   cout << "tr(B) = " << B.trace() << endl;
   cout << "tr(AB) = " << (A*B).trace() << endl;
   cout << "tr(BA) = " << (B*A).trace() << endl;
   cout << "tr(A)tr(B) = " << A.trace() * B.trace() << endl;
}
```

```
Result
======
A =
[2 -1 1]
[1 -2 -1]
[3 2 2]

B =
[-1 -1 -1]
[-1 -1 -1]
[-1 -1 -1]

tr(A) = 2
tr(B) = -3
tr(AB) = -7
tr(BA) = -7
tr(A)tr(B) = -6
```

Example 3. We demonstrate the usage of the determinant function.

```
// deter.cxx

#include <iostream.h>
#include "Matrix.h"

void main()
{
   Matrix<double> A(2,2);

   A[0][0] = 1; A[0][1] = 2;
   A[1][0] = 3; A[1][1] = 4;

   cout << A;
   cout << "Determinant of the matrix = " << A.determinant() << endl;
   cout << endl;

   for (int i=3; i<5; i++)
   {
 A.resize(i,i,i);
 cout << A;
 cout << "Determinant of the matrix = " << A.determinant() << endl;
 cout << endl;
   }
}
```

```
Result
======
[1 2]
[3 4]
Determinant of the matrix = -2

[1 2 3]
[3 4 3]
[3 3 3]
Determinant of the matrix = -6

[1 2 3 4]
[3 4 3 4]
[3 3 3 4]
[4 4 4 4]
Determinant of the matrix = 8
```

5.4 Header File Vector Class

This section contains the listing of the header files for the vector class. A brief description of the public member functions are given prior to the complete program listing.

The public interface of the `Vector` class:

- `Vector()` : Default constructor.
- `Vector(int)` : Constructor.
- `Vector(int, T)` : Constructor.
- `length()` : Length of the vector.
- `resize(int n)` : Resizes the vector to `n`.
- `resize(int n, T v)` : Resizes the vector to `n` and fills the rest of the vector entries with value `v`.
- `reset(int n)` : Resets the vector to size `n`.
- `reset(int n, T v)` : Resets the vector to size `n` and initializes entries to `v`.
- Arithmetic operators : `+`(unary), `-`(unary), `+`, `-`, `*`, `/`, `=`, `+=`, `-=`, `*=`, `/=`
- Relational operators : `==`, `!=`
- Subscript operator : `[]`
- Scalar product : `|`
- Vector product : `%`
- Stream operators : `>>`, `<<`

Auxiliary functions in `VecNorm.h`:

- `norm1(const Vector &)` : One-norm of the vector.
- `norm2(const Vector &)` : Two-norm of the vector.
- `normI(const Vector &)` : Infinite-norm of the vector.
- `normalize(const Vector &)` : Normalization of the vector.

Program Listing

```
// Vector.h
// Vector class

#ifndef MVECTOR_H
#define MVECTOR_H

#include <iostream.h>
#include <string.h>
#include <math.h>
#include <assert.h>

// forward declaration
template <class T> class Matrix;

// definition of class Vector
template <class T> class Vector
{
   private:
      // Data Fields
      int size;
      T *data;

   public:
      // Constructors
      Vector();
      Vector(int);
      Vector(int, T);
      Vector(const Vector<T> &);
      ~Vector();

      // Member Functions
      T & operator [] (int) const;

      int length() const;
      void resize(int);
      void resize(int, T);
      void reset(int);
      void reset(int, T);

      // Arithmetic Operators
      const Vector<T> & operator = (const Vector<T> &);
      const Vector<T> & operator = (T);
      Vector<T> operator + () const;
      Vector<T> operator - () const;
      Vector<T> operator += (const Vector<T> &);
      Vector<T> operator -= (const Vector<T> &);
      Vector<T> operator *= (const Vector<T> &);
      Vector<T> operator /= (const Vector<T> &);
      Vector<T> operator +  (const Vector<T> &) const;
      Vector<T> operator -  (const Vector<T> &) const;
      Vector<T> operator *  (const Vector<T> &) const;
      Vector<T> operator /  (const Vector<T> &) const;
```

```
      Vector<T> operator += (T);
      Vector<T> operator -= (T);
      Vector<T> operator *= (T);
      Vector<T> operator /= (T);
      Vector<T> operator +  (T) const;
      Vector<T> operator -  (T) const;
      Vector<T> operator *  (T) const;
      Vector<T> operator /  (T) const;
      T operator | (const Vector<T> &);
      Vector<T> operator % (const Vector<T> &);

      // I/O stream functions
      friend class Matrix<T>;
      friend ostream & operator << (ostream &, const Vector<T> &);
      friend istream & operator >> (istream &, Vector<T> &);
};

// implementation of class Vector
template <class T> Vector<T>::Vector()
   : size(0), data(NULL) {}

template <class T> Vector<T>::Vector(int n)
   : size(n), data(new T[n])
{  assert(data != NULL); }

template <class T> Vector<T>::Vector(int n, T value)
   : size(n), data(new T[n])
{
   assert(data != NULL);
   for (int i=0; i<n; i++) data[i] = value;
}

template <class T> Vector<T>::Vector(const Vector<T> &v)
   : size(v.size), data(new T[v.size])
{
   assert(data != NULL);
   for (int i=0; i<v.size; i++) data[i] = v.data[i];
}

template <class T> Vector<T>::~Vector()
{
   delete [] data;
}

template <class T> T & Vector<T>::operator[] (int i) const
{
   assert(i >= 0 && i < size);
   return data[i];
}

template <class T> int Vector<T>::length() const
{
   return size;
}
```

```
template <class T> void Vector<T>::resize(int length)
{
   int i;
   T zero(0);
   T *newData = new T[length]; assert(newData != NULL);
   if (length <= size)
      for (i=0; i<length; i++) newData[i] = data[i];
   else
   {
      for (i=0; i<size; i++)   newData[i] = data[i];

      for (i=size; i<length; i++) newData[i] = zero;
   }
   delete [] data;
   size = length;
   data = newData;
}

template <class T> void Vector<T>::resize(int length, T value)
{
   int i;
   T *newData = new T[length]; assert(newData != NULL);
   if (length <= size)
      for (i=0; i<length; i++)    newData[i] = data[i];
   else
   {
      for (i=0; i<size; i++)      newData[i] = data[i];
      for (i=size; i<length; i++) newData[i] = value;
   }
   delete [] data;
   size = length;
   data = newData;
}

template <class T> void Vector<T>::reset(int length)
{
   T zero(0);
   delete [] data;
   data = new T[length]; assert(data != NULL);
   size = length;
   for (int i=0; i<size; i++) data[i] = zero;
}

template <class T> void Vector<T>::reset(int length, T value)
{
   delete [] data;
   data = new T[length]; assert(data != NULL);
   size = length;
   for (int i=0; i<size; i++) data[i] = value;
}

template <class T>
const Vector<T> & Vector<T>::operator = (const Vector<T> &v)
```

```
{
   if (this == &v) return *this;
   if (size != v.size)
   {
      delete [] data;
      data = new T[v.size]; assert(data != NULL);
      size = v.size;
   }
   for (int i=0; i<v.size; i++) data[i] = v.data[i];
   return *this;
}

template <class T> const Vector<T> & Vector<T>::operator = (T value)
{
   for (int i=0; i<size; i++) data[i] = value;
   return *this;
}

template <class T> Vector<T> Vector<T>::operator + () const
{ return *this; }

template <class T> Vector<T> Vector<T>::operator - () const
{ return *this * T(-1); }

template <class T> Vector<T> Vector<T>::operator += (const Vector<T> &v)
{
   assert(size==v.size);
   for (int i=0; i<size; i++) data[i] += v.data[i];
   return *this;
}

template <class T> Vector<T> Vector<T>::operator -= (const Vector<T> &v)
{
   assert(size==v.size);
   for (int i=0; i<size; i++) data[i] -= v.data[i];
   return *this;
}

template <class T> Vector<T> Vector<T>::operator *= (const Vector<T> &v)
{
   assert(size==v.size);
   for (int i=0; i<size; i++) data[i] *= v.data[i];
   return *this;
}

template <class T> Vector<T> Vector<T>::operator /= (const Vector<T> &v)
{
   assert(size==v.size);
   for (int i=0; i<size; i++) data[i] /= v.data[i];
   return *this;
}

template <class T>
Vector<T> Vector<T>::operator + (const Vector<T> &v) const
```

```
{
   Vector<T> result(*this);
   return  result += v;
}

template <class T>
Vector<T> Vector<T>::operator - (const Vector<T> &v) const
{
   Vector<T> result(*this);
   return  result -= v;
}

template <class T>
Vector<T> Vector<T>::operator * (const Vector<T> &v) const
{
   Vector<T> result(*this);
   return  result *= v;
}

template <class T>
Vector<T> Vector<T>::operator / (const Vector<T> &v) const
{
   Vector<T> result(*this);
   return  result /= v;
}

template <class T> Vector<T> Vector<T>::operator += (T c)
{
   for (int i=0; i<size; i++) data[i] += c;
   return *this;
}

template <class T> Vector<T> Vector<T>::operator -= (T c)
{
   for (int i=0; i<size; i++) data[i] -= c;
   return *this;
}

template <class T> Vector<T> Vector<T>::operator *= (T c)
{
   for (int i=0; i<size; i++) data[i] *= c;
   return *this;
}

template <class T> Vector<T> Vector<T>::operator /= (T c)
{
   for (int i=0; i<size; i++) data[i] /= c;
   return *this;
}

template <class T> Vector<T> Vector<T>::operator + (T c) const
{
   Vector<T> result(*this);
   return result += c;
```

```
}

template <class T> Vector<T> Vector<T>::operator - (T c) const
{
   Vector<T> result(*this);
   return result -= c;
}

template <class T> Vector<T> Vector<T>::operator * (T c) const
{
   Vector<T> result(*this);
   return result *= c;
}

template <class T> Vector<T> Vector<T>::operator / (T c) const
{
   Vector<T> result(*this);
   return result /= c;
}

template <class T> Vector<T> operator + (T c, const Vector<T> &v)
{ return v+c; }

template <class T> Vector<T> operator - (T c, const Vector<T> &v)
{ return -v+c; }

template <class T> Vector<T> operator * (T c, const Vector<T> &v)
{ return v*c; }

template <class T> Vector<T> operator / (T c, const Vector<T> &v)
{
   Vector<T> result(v.length());
   for (int i=0; i<result.length(); i++)
      result[i] = c/v[i];
   return result;
}

// Dot Product / Inner Product
template <class T> T Vector<T>::operator | (const Vector<T> &v)
{
   assert(size == v.size);
   T result(0);
   for (int i=0; i<size; i++) result = result + data[i]*v.data[i];
   return result;
}

// Cross Product
template <class T>
Vector<T> Vector<T>::operator % (const Vector<T> &v)
{
   assert(size == 3 && v.size == 3);
   Vector<T> result(3);
   result.data[0] = data[1] * v.data[2] - v.data[1] * data[2];
   result.data[1] = v.data[0] * data[2] - data[0] * v.data[2];
```

```
   result.data[2] = data[0] * v.data[1] - v.data[0] * data[1];
   return result;
}

// Equality
template <class T>
int operator == (const Vector<T> &u, const Vector<T> &v)
{
   if (u.length() != v.length()) return 0;
   for (int i=0; i<u.length(); i++)
      if (u[i] != v[i]) return 0;
   return 1;
}

// Inequality
template <class T>
int operator != (const Vector<T> &u, const Vector<T> &v)
{
   return !(u==v);
}

template <class T> ostream & operator << (ostream &s, const Vector<T> &v)
{
   int lastnum = v.length();
   for (int i=0; i<lastnum; i++) s << "[" << v[i] << "]" << endl;
   return s;
}

template <class T> istream & operator >> (istream &s, Vector<T> &v)
{
   int i, num;
   s.clear();                  // set stream state to good
   s >> num;                   // read size of Vector
   if (! s.good()) return s;   // can't get an integer, just return
   v.resize(num);              // resize Vector v
   for (i=0; i<num; i++)
   {
      s >> v[i];               // read in entries
      if (! s.good())
      {
 s.clear(s.rdstate() | ios::badbit);
 return s;
      }
   }
   return s;
}
#endif
```

```
// VecNorm.h
// Norms of Vectors

#ifndef MVECNORM_H
#define MVECNORM_H

#include <iostream.h>
#include <math.h>

template <class T> T norm1(const Vector<T> &v)
{
   T result(0);

   for (int i=0; i<v.length(); i++)
      result = result + abs(v[i]);

   return result;
}

double norm1(const Vector<double> &v)
{
   double result(0);

   for (int i=0; i<v.length(); i++)
      result = result + fabs(v[i]);

   return result;
}

template <class T> double norm2(const Vector<T> &v)
{
   T result(0);

   for (int i=0; i<v.length(); i++)
      result = result + v[i]*v[i];

   return sqrt(double(result));
}

template <class T> T normI(const Vector<T> &v)
{
   T maxItem(abs(v[0])),
     temp;

   for (int i=1; i<v.length(); i++)
   {
      temp = abs(v[i]);
      if (temp > maxItem) maxItem = temp;
   }
   return maxItem;
}

double normI(const Vector<double> &v)
{
```

```
   double maxItem(fabs(v[0])),
  temp;

   for (int i=1; i<v.length(); i++)
   {
      temp = fabs(v[i]);
      if (temp > maxItem) maxItem = temp;
   }
   return maxItem;
}

template <class T> Vector<T> normalize(const Vector<T> &v)
{
   Vector<T> result(v.length());
   double length = norm2(v);

   for (int i=0; i<v.length(); i++)
      result[i] = v[i]/length;

   return result;
}
#endif
```

5.5 Header File Matrix Class

The public interface of the `Matrix` class:

- `Matrix()` : Default constructor.
- `Matrix(int, int)` : Constructor.
- `Matrix(int, int, T)` : Constructor.
- `Matrix(const Vector<T> &)` : Constructor.
- `identity()` : Creates an identity matrix.
- `transpose()` : Transpose of the matrix.
- `inverse()` : Inverse of the matrix.
- `LU()` : LU decomposition of the matrix.
- `trace()` : Trace of the matrix.
- `determinant()` : Determinant of the matrix.
- `rows()` : Number of rows of the matrix.
- `cols()` : Number of columns of the matrix.
- `resize(int, int)` : Resizes the matrix.
- `resize(int, int, T)` : Resizes the matrix and initializes the rest of the entries.
- `vec(const Matrix<T> &)` : Vectorize operator.
- `kron(const Matrix<T> &, const Matrix<T> &)` : Kronecker product.
- `fill(T v)` : Fills the matrix with the value `v`.
- Arithmetic operators : `+`(unary), `-`(unary), `+`, `-`, `*`, `/`, `=`, `+=`, `-=`, `*=`, `/=`
- Row vector operator : `[]`
- Column vector operator : `()`
- Stream operators : `>>`, `<<`

Auxiliary functions in `MatNorm.h`:

- `norm1(const Matrix &)` : One-norm of the matrix.
- `normI(const Matrix &)` : Infinite-norm of the matrix.
- `normH(const Matrix &)` : Hilbert-Schmidt norm of the matrix.

Program Listing

```
// Matrix.h
// Matrix class

#ifndef MATRIX_H
#define MATRIX_H

#include <iostream.h>
#include <math.h>
#include <assert.h>
#include "Vector.h"

// definition of class Matrix
template <class T> class Matrix
{
   protected:
      // Data Fields
      int rowNum, colNum;
      Vector<T> *mat;

   public:
      // Constructors
      Matrix();
      Matrix(int, int);
      Matrix(int, int, T);
      Matrix(const Vector<T> &);
      Matrix(const Matrix<T> &);
      ~Matrix();

      // Member Functions
      Vector<T> & operator [] (int) const;
      Vector<T>   operator () (int) const;

      Matrix<T> identity();
      Matrix<T> transpose() const;
      Matrix<T> inverse() const;
      T trace() const;
      T determinant() const;

      int rows() const;
      int cols() const;
      void resize(int, int);
      void resize(int, int, T);
      void fill(T);

      // Arithmetic Operators
      const Matrix<T> & operator = (const Matrix<T> &);
      const Matrix<T> & operator = (T);

      Matrix<T> operator + () const;
      Matrix<T> operator - () const;
      Matrix<T> operator += (const Matrix<T> &);
      Matrix<T> operator -= (const Matrix<T> &);
```

```
      Matrix<T> operator *= (const Matrix<T> &);
      Matrix<T> operator +  (const Matrix<T> &) const;
      Matrix<T> operator -  (const Matrix<T> &) const;
      Matrix<T> operator *  (const Matrix<T> &) const;
      Vector<T> operator *  (const Vector<T> &) const;

      Matrix<T> operator += (T);
      Matrix<T> operator -= (T);
      Matrix<T> operator *= (T);
      Matrix<T> operator /= (T);
      Matrix<T> operator +  (T) const;
      Matrix<T> operator -  (T) const;
      Matrix<T> operator *  (T) const;
      Matrix<T> operator /  (T) const;

      friend Vector<T> vec(const Matrix<T> &);
      friend Matrix<T> kron(const Matrix<T> &, const Matrix<T> &);
      friend ostream & operator << (ostream &, const Matrix<T>&);
      friend istream & operator >> (istream &, Matrix<T>&);
};

// implementation of class Matrix
template <class T> Matrix<T>::Matrix()
   : rowNum(0), colNum(0), mat(NULL) {}

template <class T> Matrix<T>::Matrix(int r, int c)
   : rowNum(r), colNum(c), mat(new Vector<T>[r])
{
   assert(mat != NULL);
   for (int i=0; i<r; i++) mat[i].resize(c);
}

template <class T> Matrix<T>::Matrix(int r, int c, T value)
   : rowNum(r), colNum(c), mat(new Vector<T>[r])
{
   assert(mat != NULL);
   for (int i=0; i<r; i++) mat[i].resize(c,value);
}

template <class T> Matrix<T>::Matrix(const Vector<T> &v)
   : rowNum(v.length()), colNum(1), mat(new Vector<T>[rowNum])
{
   assert(mat != NULL);
   for (int i=0; i<rowNum; i++) mat[i].resize(1,v[i]);
}

template <class T> Matrix<T>::Matrix(const Matrix<T> &m)
   : rowNum(m.rowNum), colNum(m.colNum), mat(new Vector<T>[m.rowNum])
{
   assert(mat != NULL);
   for (int i=0; i<m.rowNum; i++) mat[i] = m.mat[i];
}

template <class T> Matrix<T>::~Matrix()
```

```
{
   delete [] mat;
}

template <class T> Vector<T> & Matrix<T>::operator [] (int index) const
{
   assert(index>=0 && index<rowNum);
   return mat[index];
}

template <class T> Vector<T> Matrix<T>::operator () (int index) const
{
   assert(index>=0 && index<colNum);
   Vector<T> result(rowNum);

   for (int i=0; i<rowNum; i++) result[i] = mat[i][index];

   return result;
}

template <class T> Matrix<T> Matrix<T>::identity()
{
   for (int i=0; i<rowNum; i++)
      for (int j=0; j<colNum; j++)
 if (i==j) mat[i][j] = T(1);
         else      mat[i][j] = T(0);
   return *this;
}

template <class T> Matrix<T> Matrix<T>::transpose() const
{
   Matrix<T> result(colNum,rowNum);

   for (int i=0; i<rowNum; i++)
      for (int j=0; j<colNum; j++)
         result[j][i] = mat[i][j];

   return result;
}

// Symbolical Inverse using Leverrier's Method
template <class T> Matrix<T> Matrix<T>::inverse() const
{
   assert(rowNum == colNum);
   Matrix<T> B(*this), D, I(rowNum, colNum);
   T c0(B.trace()), c1;
   int i;

   I.identity();

   for (i=2; i<rowNum; i++)
   {
      B = *this * (B-c0*I);
      c0 = B.trace()/T(i);
```

```
   }
   D = *this * (B-c0*I);
   c1 = D.trace()/T(i);

   return (B-c0*I)/c1;
}

template <class T> T Matrix<T>::trace() const
{
   assert(rowNum == colNum);
   T result(0);

   for (int i=0; i<rowNum; i++) result += mat[i][i];

   return result;
}

// Symbolical determinant
template <class T> T Matrix<T>::determinant() const
{
   assert(rowNum==colNum);
   Matrix<T> B(*this), I(rowNum, colNum, T(0));
   T c(B.trace());
   int i;

   for (i=0; i<rowNum; i++) I[i][i] = T(1);

   // Note that determinant of int-type gives zero
   // because of division by T(i)
   for (i=2; i<=rowNum; i++)
   {
      B = *this * (B-c*I);
      c = B.trace()/T(i);
   }

   if (rowNum%2) return c;
   return -c;
}

template <class T> int Matrix<T>::rows() const
{
   return rowNum;
}

template <class T> int Matrix<T>::cols() const
{
   return colNum;
}

template <class T> void Matrix<T>::resize(int r, int c)
{
   int i;
   Vector<T> *newMat = new Vector<T>[r]; assert(newMat != NULL);
```

```
   if (r<=rowNum)
   {
      for (i=0; i<r; i++)
      {
         (mat+i) -> resize(c);
         newMat[i] = mat[i];
      }
   }
   else
   {
      for (i=0; i<rowNum; i++)
      {
         (mat+i) -> resize(c);
         newMat[i] = mat[i];
      }
      for (i=rowNum; i<r; i++) newMat[i].resize(c);
   }
   delete [] mat;

   rowNum = r; colNum = c;
   mat = newMat;
}

template <class T> void Matrix<T>::resize(int r, int c, T value)
{
   int i;
   Vector<T> *newMat = new Vector<T>[r]; assert(newMat != NULL);

   if (r<=rowNum)
   {
      for (i=0; i<r; i++)
      {
         (mat+i) -> resize(c,value);
 newMat[i] = mat[i];
      }
   }
   else
   {
      for (i=0; i<rowNum; i++)
      {
 (mat+i) -> resize(c,value);
         newMat[i] = mat[i];
      }
      for (i=rowNum; i<r; i++) newMat[i].resize(c,value);
   }
   delete [] mat;

   rowNum = r; colNum = c;
   mat = newMat;
}

template <class T> void Matrix<T>::fill(T value)
{
   for (int i=0; i<rowNum; i++)
```

```
      for (int j=0; j<colNum; j++)
         mat[i][j] = value;
}

template <class T>
const Matrix<T> & Matrix<T>::operator = (const Matrix<T> &m)
{
   if (this == &m) return *this;

   delete [] mat;
   rowNum = m.rowNum; colNum = m.colNum;

   mat = new Vector<T>[m.rowNum]; assert(mat != NULL);

   for (int i=0; i<m.rowNum; i++) mat[i] = m.mat[i];

   return *this;
}

template <class T>
const Matrix<T> & Matrix<T>::operator = (T value)
{
   for (int i=0; i<rowNum; i++) mat[i] = value;
   return *this;
}

template <class T> Matrix<T> Matrix<T>::operator + () const
{
   return *this;
}

template <class T> Matrix<T> Matrix<T>::operator - () const
{
   return *this * T(-1);
}

template <class T> Matrix<T> Matrix<T>::operator += (const Matrix<T> &m)
{
   return *this = *this + m;
}

template <class T> Matrix<T> Matrix<T>::operator -= (const Matrix<T> &m)
{
   return *this = *this - m;
}

template <class T> Matrix<T> Matrix<T>::operator *= (const Matrix<T> &m)
{
   return *this = *this * m;
}

template <class T>
Matrix<T> Matrix<T>::operator + (const Matrix<T> &m) const
{
```

```
   assert(rowNum == m.rowNum && colNum == m.colNum);

   Matrix<T> result(*this);

   for (int i=0; i<rowNum; i++) result[i] += m[i];
   return result;
}

template <class T>
Matrix<T> Matrix<T>::operator - (const Matrix<T> &m) const
{
   assert(rowNum == m.rowNum && colNum == m.colNum);

   Matrix<T> result(*this);

   for (int i=0; i<rowNum; i++) result[i] -= m[i];
   return result;
}

template <class T>
Matrix<T> Matrix<T>::operator * (const Matrix<T> &m) const
{
   assert(colNum == m.rowNum);

   Matrix<T> result(rowNum, m.colNum, T(0));

   for (int i=0; i<rowNum; i++)
      for (int j=0; j<m.colNum; j++)
         for (int k=0; k<colNum; k++)
    result[i][j] += mat[i][k] * m[k][j];
   return result;
}

template <class T>
Vector<T> Matrix<T>::operator * (const Vector<T> &v) const
{
   assert(colNum == v.length());

   Vector<T> result(rowNum);

   // dot product | is used
   for (int i=0; i<rowNum; i++) result[i] = (mat[i] | v);

   return result;
}

template <class T> Matrix<T> Matrix<T>::operator += (T c)
{
   assert(rowNum == colNum);
   for (int i=0; i<rowNum; i++) mat[i][i] += c;
   return *this;
}

template <class T> Matrix<T> Matrix<T>::operator -= (T c)
```

```
{
   assert(rowNum == colNum);
   for (int i=0; i<rowNum; i++) mat[i][i] -= c;
   return *this;
}

template <class T> Matrix<T> Matrix<T>::operator *= (T c)
{
   for (int i=0; i<rowNum; i++) mat[i] *= c;
   return *this;
}

template <class T> Matrix<T> Matrix<T>::operator /= (T c)
{
   for (int i=0; i<rowNum; i++) mat[i] /= c;
   return *this;
}

template <class T>
Matrix<T> Matrix<T>::operator + (T value) const
{
   assert(rowNum == colNum);
   Matrix<T> result(*this);
   return result += value;
}

template <class T>
Matrix<T> Matrix<T>::operator - (T value) const
{
   assert(rowNum == colNum);
   Matrix<T> result(*this);
   return result -= value;
}

template <class T>
Matrix<T> Matrix<T>::operator * (T value) const
{
   Matrix<T> result(*this);
   return result *= value;
}

template <class T>
Matrix<T> Matrix<T>::operator / (T value) const
{
   Matrix<T> result(*this);
   return result /= value;
}

template <class T>
Matrix<T> operator + (T value, const Matrix<T> &m)
{
   return m + value;
}
```

```
template <class T>
Matrix<T> operator - (T value, const Matrix<T> &m)
{
   return -m + value;
}

template <class T>
Matrix<T> operator * (T value, const Matrix<T> &m)
{
   return m * value;
}

template <class T>
Matrix<T> operator / (T value, const Matrix<T> &m)
{
   Matrix<T> result(m.rows(),m.cols());

   for (int i=0; i<result.rows(); i++) result[i] = value/m[i];
   return result;
}

// Vectorize operator
template <class T> Vector<T> vec(const Matrix<T> &m)
{
   int i=0, j, k, size = m.rowNum * m.colNum;
   Vector<T> result(size);

   for (j=0; j<m.colNum; j++)
      for (k=0; k<m.rowNum; k++) result[i++] = m.mat[k][j];
   return result;
}

// Kronecker Product
template <class T>
Matrix<T> kron(const Matrix<T> &s, const Matrix<T> &m)
{
   int size1 = s.rowNum * m.rowNum,
       size2 = s.colNum * m.colNum,
       i, j, k, p;
   Matrix<T> result(size1, size2);

   for (i=0; i<s.rowNum; i++)
      for (j=0; j<s.colNum; j++)
         for (k=0; k<m.rowNum; k++)
            for (p=0; p<m.colNum; p++)
        result[k + i*m.rowNum][p + j*m.colNum]
                 = s.mat[i][j] * m.mat[k][p];
   return result;
}

template <class T>
int operator == (const Matrix<T> &m1, const Matrix<T> &m2)
{
   if (m1.rows() != m2.rows()) return 0;
```

```
   for (int i=0; i<m1.rows(); i++)
      if (m1[i] != m2[i]) return 0;

   return 1;
}

template <class T>
int operator != (const Matrix<T> &m1, const Matrix<T> &m2)
{
   return !(m1==m2);
}

template <class T> ostream & operator << (ostream &s, const Matrix<T> &m)
{
   int t = m.cols()-1;

   for (int i=0; i<m.rows(); i++)
   {
      s << "[";
      for (int j=0; j<t; j++) s << m[i][j] << " ";
      s << m[i][t] << "]" << endl;
   }

   return s;
}

template <class T> istream & operator >> (istream &s, Matrix<T> &m)
{
   int i, j, num1, num2;

   s.clear();                 // set stream state to good
   s >> num1;                 // read in row number
   if (! s.good()) return s;  // can't get an integer, just return

   s >> num2;                 // read in column number
   if (! s.good()) return s;  // can't get an integer, just return

   m.resize(num1,num2);       // resize to Matrix into right order

   for (i=0; i<num1; i++)
      for (j=0; j<num2; j++)
      {
 s >> m[i][j];
 if (! s.good())
 {
    s.clear(s.rdstate() | ios::badbit);
    return s;
 }
      }
   return s;
}
#endif
```

```
// MatNorm.h
// Norms of Matrices

#ifndef MATNORM_H
#define MATNORM_H

#include <iostream.h>
#include <math.h>
#include "Vector.h"
#include "VecNorm.h"

template <class T> T norm1(const Matrix<T> &m)
{
   T maxItem(0), temp;
   int i,j;

   for (i=0; i<m.rows(); i++)
      maxItem += m[i][0];

   for (i=1; i<m.cols(); i++)
   {
      temp = T(0);
      for (j=0; j<m.rows(); j++)
 temp += abs(m[j][i]);

      if (temp > maxItem) maxItem = temp;
   }
   return maxItem;
}

template <class T> T normI(const Matrix<T> &m)
{
   T maxItem(norm1(m[0]));

   for (int i=1; i<m.rows(); i++)
      if (norm1(m[i]) > maxItem) maxItem = norm1(m[i]);

   return maxItem;
}

template <class T> T normH(const Matrix<T> &m)
{
   return sqrt((m*(m.transpose())).trace());
}
#endif
```

Bibliography

[1] Abe, Eiichi *Hopf Algebras*, Cambridge University Press, 1977, Cambridge.

[2] Angle, E. and Bellman, R. *Dynamic Programming and Partial Differential Equations*, Academic Press, 1972, New York.

[3] Barnett, S. *Matrix Differential Equations And Kronecker Products*, SIAM J. Appl. Math. 1973, **24**, 1–5.

[4] Barnett, S. *Inversion of partitioned matrices with patterned blocks*, Int. J. Systems Sci. 1983, **14**, 235–237.

[5] Barouch, E. *Lax Pair for the Free Fermion Eight-Vertex Model*, Studies in Applied Mathematics 1982, **70**, 151–162.

[6] Baumslag B. and Chandler, B. *Group Theory*, Schaum's Outline Series, McGraw-Hill, 1968, New York.

[7] Baxter, R. J. *Exactly Solved Models in Statistical Mechanics*, Academic Press, 1982, New York.

[8] Brewer, J. W. *Kronecker Products and Matrix Calculus in System Theory*, IEEE Transactions on Circuits and Systems, 1978, CAS. 25, 772 – 781.

[9] Ciarlet, P. G. *Introduction to numerical linear algebra and optimisation*, Cambridge University Press, 1989.

[10] Davis, P. J. *Circulant Matrices*, John Wiley, 1979, New York.

[11] Deif, A. S. *Advanced Matrix Theory for Scientists and Engineers*, Abacus Press-Halsted Press, 1982, London.

[12] Elby, A. and Bub J. *Triorthogonal uniqueness theorem and its relevance to the interpretation of quantum mechanics*, Phys. Rev. A 1994, **49**, 4213–4216.

[13] Elliott, D. F. and Rao, K. R. *Fast Fourier Transforms*, Academic Press, 1982, Orlando.

[14] Graham A. *Kronecker Product and Matrix Calculus with Applications*, Ellis Horwood Limited, Chichester, 1981.

[15] Gröbner, W. *Matrizenrechnung*, Bibliographisches Institut, 1966, Mannheim.

[16] Huang, K. *Statistical Mechanics*, Second Edition, John Wiley, 1987, New York.

[17] Humphreys, J. E. *Introduction to Lie Algebras and Representation Theory*, Springer-Verlag 1980, New York.

[18] Kaufman, B. *Crystal Statistic. II. Partition Function Evaluated by Spinor Analysis*, Physical Review 1949, **76**, 1232–1243.

[19] Kent, A. *A note on Schmidt states and consistency*, Phys. Lett. A 1995, **196**, 313–317.

[20] Kulish, P. P. and Sklyanin, E. K. *Quantum Inverse Scattering Method and the Heisenberg Ferromagnet*, Physics Letters 1979, **70A**, 461–463.

[21] Lancaster, P. *Theory of Matrices* Academic Press, 1969, New York.

[22] Lieb, E. H. *Solution of the Dimer Problem by the Transfer Matrix Method*, Journal of Mathematical Physics 1967, **8**, 2339–2341.

[23] Ludwig, W. and Falter, C. *Symmetries in Physics*, Springer, 1988, Berlin.

[24] Miller, W. *Symmetry Groups and Their Applications*, Academic Press, 1972, New York.

[25] Neudecker, H. *Some Theorems on matrix differentiation with special reference to Kronecker matrix products*, Journal of the American Statistical Association 1969, **64**, 953–963.

[26] Onsager, L. *Zero Field Properties of the Plane Square Lattice*, Physical Review 1944, **65**, 117–149.

[27] Percus, J. K. *Combinatorical Methods*, Springer, 1971, New York.

[28] Prugovečki, E. *Quantum Mechanics in Hilbert Space*, Second Edition, Academic Press, 1981, New York.

[29] Regalia, P. A. and Mitra, S. K. *Kronecker products, unitary matrices and signal processing applications*, SIAM Review 1989, 586 – 613.

[30] Searle, S. R. *Matrix Algebra Useful For Statistics*, Wiley, 1982, New York.

[31] Sogo, K. and Wadati, M. *Boost Operator and Its Application to Quantum Gelfand-Levitan Equation for Heisenberg-Ising Chain with Spin One-Half*, Progress of Theoretical Physics 1983, **69**, 431–450.

[32] Steeb, W.-H. *A Comment on Trace Calculations for Fermi Systems*, Acta Physica Adacemiae Scientiarum Hungaricae 1977, **42**, 171 – 177.

[33] Steeb, W.-H. *The Relation between the Kronecker Product, the Trace Calculation, and the One-dimensional Ising Model*, Linear Algebra and its Application 1981, **37**, 261 – 265.

[34] Steeb, W.-H. and Wilhelm, F. *Exponential Functions of Kronecker Products and Trace Calculation*, Linear and Multilinear Algebra 1981, **9**, 345 – 346.

[35] Steeb, W.-H. and Solms, F. *Applications of C++ Programming*, World Scientific, Singapore 1995.

[36] Steeb, W.-H. *Continuous Symmetries, Lie Algebras, Differential Equations and Computer Algebra*, World Scientific, Singapore 1996.

[37] Steeb, W.-H. *Problems in Theoretical and Mathematical Physics, Volume I, Introductory Level*, World Scientific, Singapore 1996.

[38] Steeb, W.-H. *Problems in Theoretical and Mathematical Physics, Volume II, Advanced Level*, World Scientific, Singapore 1996.

[39] Steeb, W.-H. and Lai Choy Heng *Lax Representation and Kronecker Product*, Int. J. Theor. Phys. 1996, **35**, 475–479.

[40] Steiner, P. A. J. *Real Clifford algebras and their representations over the reals*, Journal of Physics A: Math. Gen. 1987, **20**, 3095 – 3098.

[41] Takhtadzhyan, L. A. and Faddeev, L. D. *The Quantum Method of the Inverse Problem and the Heisenberg XYZ-Model*, Russian Math. Surveys 1979, **34**, 11–68.

[42] Villet, C.M. and Steeb, W.-H. *Kronecker Products of Matrices and an Application to Fermi Systems*, International Journal of Theoretical Physics 1986, **25**, 67 – 72.

[43] Zachos, C. *Elementary Paradigms of Quantum Algebras*, Contemporary Mathematics 1992, **134**, 351–377.

Index